Cisco IOS Switching Services

Cisco Systems, Inc.

Macmillan Technical Publishing
201 West 103rd Street
Indianapolis, IN 46290 USA

Cisco IOS Switching Services

Cisco Systems, Inc.

Published by:
Macmillan Technical Publishing
201 W. 103rd Street
Indianapolis, IN 46290 USA

Printed in the United States of America 1 2 3 4 5 6 7 8 9 0

Library of Congress Cataloging-in-Publication Number 97-80673

ISBN: 1-57870-053-1

Warning and Disclaimer

This book is designed to provide information about ***Cisco IOS Switching Services.*** Every effort has been made to make this book as complete and as accurate as possible, but no warranty or fitness is implied.

The information is provided on an "as is" basis. The author, Macmillan Technical Publishing, and Cisco Systems, Inc. shall have neither liability nor responsibility to any person or entity with respect to any loss or damages arising from the information contained in this book or from the use of the discs or programs that may accompany it.

The opinions expressed in this book belong to the author and are not necessarily those of Cisco Systems, Inc.

Associate Publisher	Jim LeValley
Executive Editor	Julie Fairweather
Cisco Systems Program Manager	H. Kim Lew
Director of Editorial Services	Carla Hall
Managing Editor	Caroline Roop
Acquisitions Editor	Tracy Hughes
Development and Copy Editor	Jill Bond
Project Editor	Sherri Fugit
Acquisitions Coordinator	Amy Lewis
Manufacturing Coordinator	Brook Farling
Book Designer	Louisa Klucznik
Cover Designer	Jean Bisesi
Cover Production	Casey Price
Production Team Supervisor	Vic Peterson
Graphics Image Specialists	Patricia Douglas Laura Robbins
Production Analysts	Dan Harris Erich J. Richter
Production Team	Pamela Woolf Trina Wurst
Indexer	Tim Wright

Trademark Acknowledgments

Contents at a Glance

Table of Contents

About the Cisco IOS Reference Library

The Cisco IOS Reference Library books are Cisco documentation that describe the tasks and commands necessary to configure and to maintain your Cisco IOS network.

The Cisco IOS software bookset is intended primarily for users who configure and maintain access servers and routers, but are not necessarily familiar with the tasks, the relationship between tasks, or the commands necessary to perform particular tasks.

Cisco IOS Reference Library Organization

The Cisco IOS Reference library consists of eight books. Each book contains technology-specific configuration chapters with corresponding command reference chapters. Each configuration chapter describes Cisco's implementation of protocols and technologies, related configuration tasks, and contains comprehensive configuration examples. Each command reference chapter complements the organization of its corresponding configuration chapter and provides complete command syntax information.

Cisco IOS Switching Services Overview

The *Cisco IOS Switching Services Configuration Guide* provides guidelines for configuring switching paths and routing between virtual local area networks (VLANs) with the Cisco IOS software.

This guide is intended for the network administrator who designs and implements router-based internetworks and needs to incorporate switching, NetFlow accounting, or routing between virtual local area networks (VLANS) into the network. It presents a set of general guidelines for configuring switching of various protocols, NetFlow accounting, routing between VLANs, and local area network (LAN) emulation. The objective of this guide is to provide you with the information you need to configure any of these features.

You should know how to configure a Cisco router and should be familiar with the protocols and media that your routers are configured to support. Knowledge of basic network topology is essential.

Document Organization

This document comprises three parts, each focusing on a different aspect of switching within Cisco IOS software. Each part begins with a brief technology overview and follows with the corresponding configuration guidelines for that technology or set of features. This document contains these parts:

- **Cisco IOS Switching Paths**: Provides an overview of basic routing and switching processes. It describes switching paths available in Cisco IOS software. Configuration guidelines are provided for configuring and managing fast switching and optimum switching of various protocols.
- **NetFlow Switching**: Provides an overview of the NetFlow switching technology and describes the NetFlow accounting features. Guidelines for configuring and managing NetFlow switching are provided.
- **Virtual LANs**: Provides an overview of VLANs. Guidelines for configuring routing between VLANs using the Inter-Switch Link (ISL) and IEEE 802.10 protocols for packet encapsulation follow the overview. LAN Emulation for defining VLANs in ATM networks is described along with related configuration guidelines.

OTHER BOOKS AVAILABLE IN THE CISCO IOS REFERENCE LIBRARY

- *Cisco IOS Configuration Fundamentals*, 1-57870-044-2; Currently available

  This comprehensive guide details Cisco IOS software configuration basics. *Cisco IOS Configuration Fundamentals* offers thorough coverage of router and access server configuration and maintenance techniques. In addition to hands-on implementation and task instruction, this book also presents the complete syntax for router and access server commands, and individual examples for each command. Learn to configure interfaces in addition to system management, file loading, AutoInstall, and set up functions.

- *Cisco IOS Dial Solutions*, 1-57870-055-8; Currently available

  This book provides readers with real-world solutions and how to implement them on a network. Customers interested in implementing dial solutions across their network environment include remote sites dialing in to a central office, Internet Service Providers (ISPs), ISP customers at home offices, and enterprise WAN system administrators implementing dial-on-demand routing (DDR).

- *Cisco IOS Wide Area Networking Solutions*, 1-57870-054-x; Currently available

  This book offers thorough, comprehensive coverage of internetworking technologies, particularly ATM, Frame Relay, SMDS, LAPB, and X.25, teaching the reader how to configure the technologies in a LAN/WAN environment.

- *Cisco IOS Solutions for Network Protocols, Vol. I, IP*, 1-57870-049-3; Available April 1998

  This book is a comprehensive guide detailing available IP and IP routing alternatives. It describes how to implement IP addressing and IP services and how to configure support for a wide range of IP routing protocols, including BGP for ISP networks as well as basic and advanced IP Multicast functionality.

- *Cisco IOS Solutions for Network Protocols, Vol. II, IPX, AppleTalk and more,* 1-57870-050-7, Available April 1998

This book is a comprehensive guide detailing available network protocol alternatives. It describes how to implement various protocols in your network. This book includes documentation of the latest functionality for the IPX and AppleTalk desktop protocols as well as the following network protocols: Apollo Domain, Banyan VINES, DECNet, ISO CLNS, and XNS.

- *Cisco IOS Bridging and IBM Network Solutions*, 1-57870-051-5, Available April 1998

  This book describes Cisco's support for networks in IBM and bridging environments. Support includes: transparent and source-route transparent bridging, source-route bridging (SRB), remote source-route bridging (RSRB), data link switching plus (DLS+), serial tunnel and block serial tunnel, SDLC and LLC2 parameters, IBM network media translation, downstream physical unit and SNA service point, SNA Frame Relay access support, Advanced Peer-to-Peer Networking, and native client interface architecture (NCIA).

- *Cisco IOS Network Security*, 1-57870-057-4, Available May 1998

  This book documents security configuration from a remote site and for a central enterprise or service provider network. It describes AAA, Radius, TACACS+, and Kerberos network security features. It also explains how to encrypt data across enterprise networks. The book includes many illustrations that show configurations and functionality, along with a discussion of network security policy choices and some decision-making guidelines.

BOOK CONVENTIONS

Software and hardware documentation uses the following conventions:

- The caret character (^) represents the Control key.

  For example, the key combinations ^D and Ctrl-D are equivalent: Both mean hold down the Control key while you press the D key. Keys are indicated in capitals, but are not case-sensitive.

- A string is defined as a nonquoted set of characters.

For example, when setting an SNMP community string to *public*, do not use quotation marks around the string; otherwise, the string will include the quotation marks.

Command descriptions use these conventions:

- Vertical bars (|) separate alternative, mutually exclusive, elements.
- Square brackets ([ ]) indicate optional elements.
- Braces ({ }) indicate a required choice.
- Braces within square brackets ([{ }]) indicate a required choice within an optional element.
- **Boldface** indicates commands and keywords that are entered literally as shown.
- *Italics* indicate arguments for which you supply values; in contexts that do not allow italics, arguments are enclosed in angle brackets (< >).

Examples use these conventions:

- Examples that contain system prompts denote interactive sessions, indicating that the user enters commands at the prompt. The system prompt indicates the current command mode. For example, the prompt `Router(config)#` indicates global configuration mode.
- Terminal sessions and information the system displays are in `screen` font.
- Information you enter is in **`boldface screen`** font.
- Nonprinting characters, such as passwords, are in angle brackets (< >).
- Default responses to system prompts are in square brackets ([ ]).
- Exclamation points (!) at the beginning of a line indicate a comment line. They are also displayed by the Cisco IOS software for certain processes.

CAUTION

Means *reader be careful.* In this situation, you might do something that could result in equipment damage or loss of data.

NOTES

Means *reader take note.* Notes contain helpful suggestions or references to materials not contained in this manual.

TIMESAVER

Means *the described action saves time.* You can save time by performing the action described in the paragraph.

Within the Cisco IOS Reference Library, the term *router* is used to refer to both access servers and routers. When a feature is supported on the access server only, the term *access server* is used. When a feature is supported on one or more specific router platforms (such as the Cisco 4500), but not on other platforms (such as the Cisco 2500), the text specifies the supported platforms.

Within examples, routers and access servers are alternately shown. These products are used only for example purposes—an example that shows one product does not indicate that the other product is not supported.

PART 1

Cisco IOS Switching Paths

CHAPTER 1

Configuring Switching Paths

This chapter describes switching paths that can be configured on Cisco IOS devices. It provides an overview of switching methods, configuration guidelines for switching paths, and tuning guidelines. For documentation of other commands, use the master indexes or search online.

Overview of Basic Router Platform Architecture and Processes

To understand how switching works, it helps first to understand the basic router architecture and where various processes occur in the router.

Fast switching is enabled by default on all interfaces that support fast switching. If you have a situation in which you need to disable fast switching and fall back to the process-switching path, understanding how various processes affect the router and where they occur will help you determine your alternatives. This is especially true when you are troubleshooting traffic problems or need to process packets that require special handling. Some diagnostic or control resources are not compatible with fast switching or come at the expense of processing and switching efficiency. Understanding the effects of those resources can help you minimize their effect on network performance.

Figure 1–1 illustrates a possible internal configuration of a Cisco 7500 series router. In this configuration, the Cisco 7500 series router has an integrated Route/Switch Processor (RSP) and uses *route caching* to forward packets. The Cisco 7500 series router also uses Versatile Interface Processors (VIP). This RISC-based interface processor receives and caches routing information from the RSP. Using the routing cache, the VIP card makes switching decisions locally, relieving the RSP of involvement and speeding overall throughput. This type of switching is called *distributed switching*. Multiple VIP cards can be installed in one router.

Figure 1–1
Basic Router Architecture

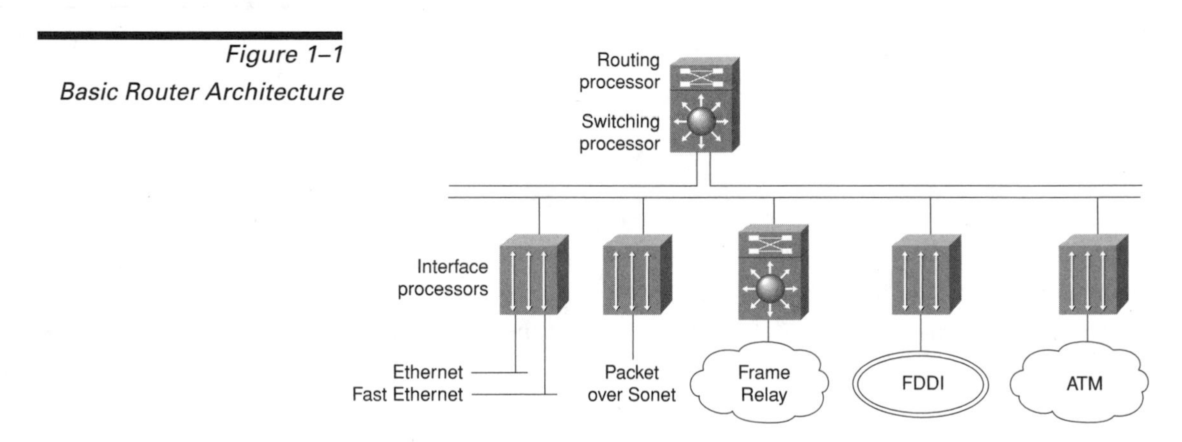

Cisco Routing and Switching Processes

The routing, or forwarding function, which comprises two interrelated processes, moves information in the network.

- Making a routing decision by routing
- Moving packets to the next-hop destination by switching

Cisco IOS platforms perform both routing and switching, and there are several types of each.

Routing

The routing process assesses the source and destination of traffic based on knowledge of network conditions. Routing functions identify the best path to use for moving the traffic to the destination, out one or more of the router interfaces. The routing decision is based upon a variety of criteria, such as link speed, topological distance, and protocol. Each separate protocol maintains its own routing information.

Routing is more processing intensive and has higher latency than switching, because it determines path and next-hop considerations. The first packet routed requires a lookup in the routing table to determine the route. The route cache is populated after the first packet is routed by the route-table lookup. Subsequent traffic for the same destination is switched using the routing information stored in the route cache. Figure 1–2 illustrates the basic routing process.

Figure 1–2
The Routing Process

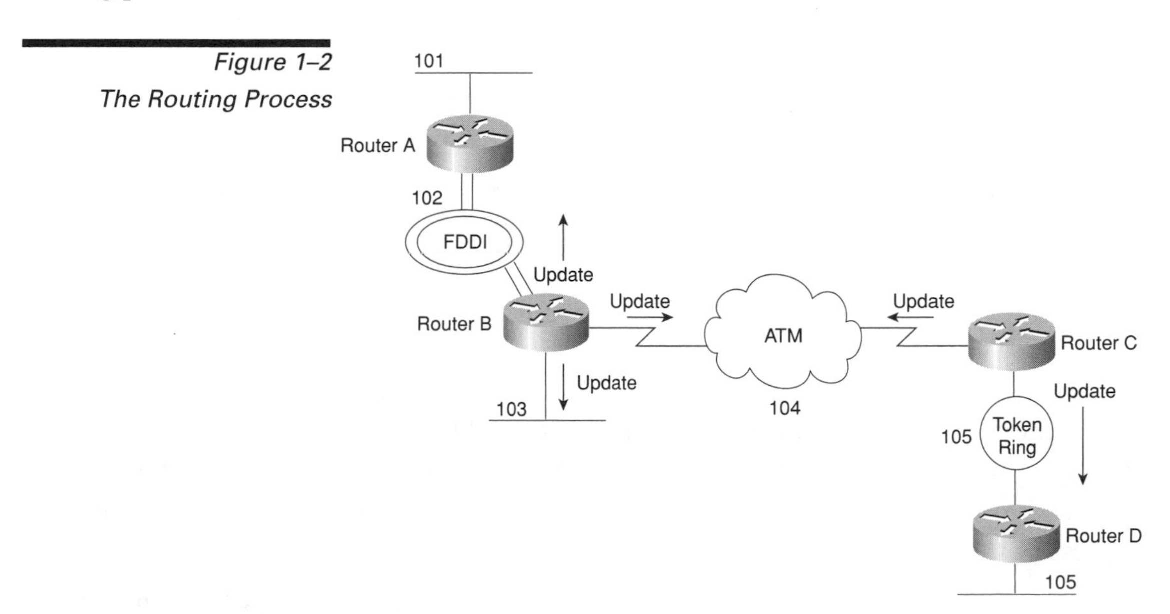

A router sends routing updates out each of its interfaces that are configured for a particular protocol. It also receives routing updates from other attached routers. From these received updates and its knowledge of attached networks, it builds a map of the network topology.

Switching

Through the switching process, the router determines the next hop toward the destination address. Switching moves traffic from an input interface to one or more output interfaces.

Switching is optimized and has lower latency than routing because it can move packets, frames, or cells from buffer to buffer with a simpler determination of the source and destination of the traffic. It saves resources because it does not involve extra lookups. Figure 1–3 illustrates the basic switching process.

Figure 1–3
The Switching Process

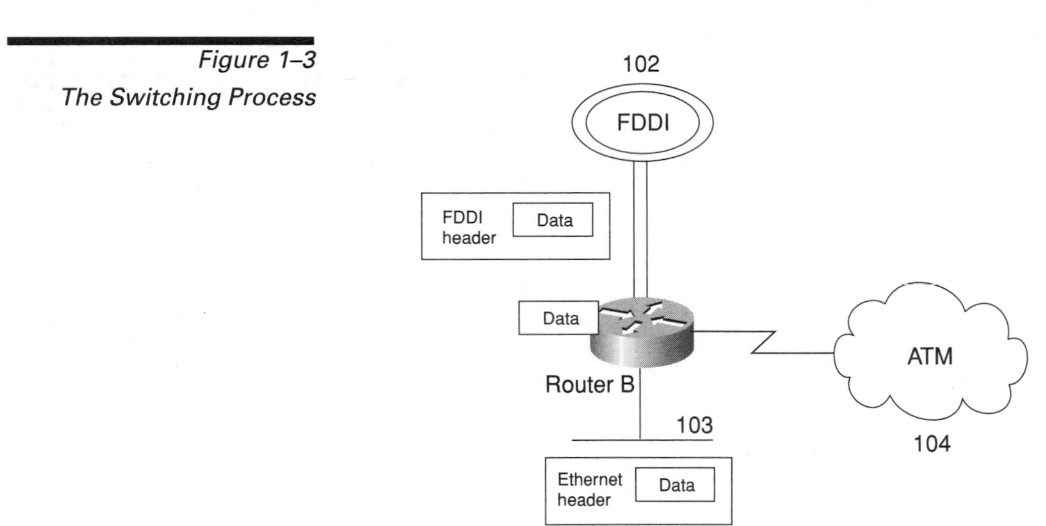

In Figure 1–3, packets are received on the Fast Ethernet interface and destined for the FDDI interface. Based on information in the packet header and destination information stored in the routing table, the router determines the destination interface. It looks in the protocol's routing table to discover the destination interface that services the destination address of the packet.

The destination address is stored in tables, such as ARP tables for IP and AARP table for AppleTalk. If there is no entry for the destination, the router must either drop the packet (and inform the user if the protocol provides that feature) or discover the destination address by some other address resolution process, such as through the ARP protocol.

Layer 3 IP addressing information is mapped to the Layer 2 MAC address for the next hop. Figure 1–4 illustrates the mapping that occurs to determine the next hop.

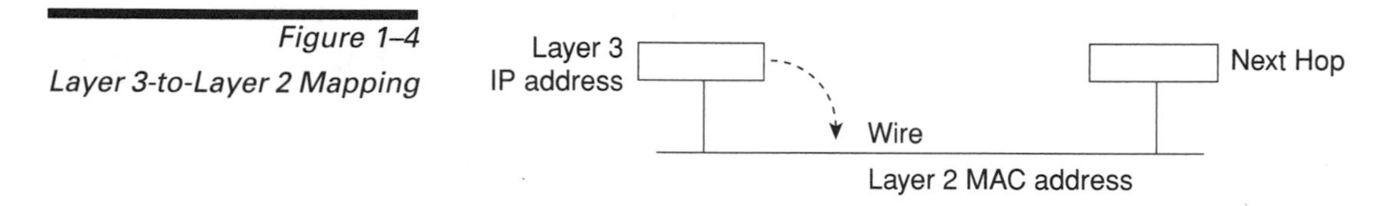

Figure 1–4
Layer 3-to-Layer 2 Mapping

Basic Switching Paths

Following are the basic switching paths:

- Process Switching
- Fast Switching
- Optimum Switching
- Distributed Switching
- NetFlow Switching

Process Switching

In process switching, the first packet is copied to the system buffer. The router looks up the Layer 3 network address in the routing table and initializes the fast-switch cache. The frame is rewritten with the destination address and sent to the exit interface that services that destination. Subsequent packets for that destination are sent by the same switching path. The route processor computes the cyclical redundancy check (CRC).

Fast Switching

When packets are fast switched, the first packet is copied to packet memory and the destination network or host is found in the fast-switching cache. The frame is rewritten and sent to the exit interface that services the destination. Subsequent packets for the same destination use the same switching path. The interface processor computes the CRC.

Optimum Switching

Optimum switching is similar to fast switching but is faster. The first packet is copied to packet memory, and the destination network or host is found in the optimum-switching cache. The frame is rewritten and sent to the exit interface that services the destination. Subsequent packets for the same destination use the same switching path. The interface processor computes the CRC.

NOTES

Optimum switching is enabled by default on Cisco 7500 series routers; it must be disabled for debugging.

Distributed Switching

The closer to the interface the function occurs, the more efficient switching becomes. In distributed switching, the switching process occurs on VIP and other interface cards that support switching. Figure 1–5 illustrates the distributed switching process on the Cisco 7500 series.

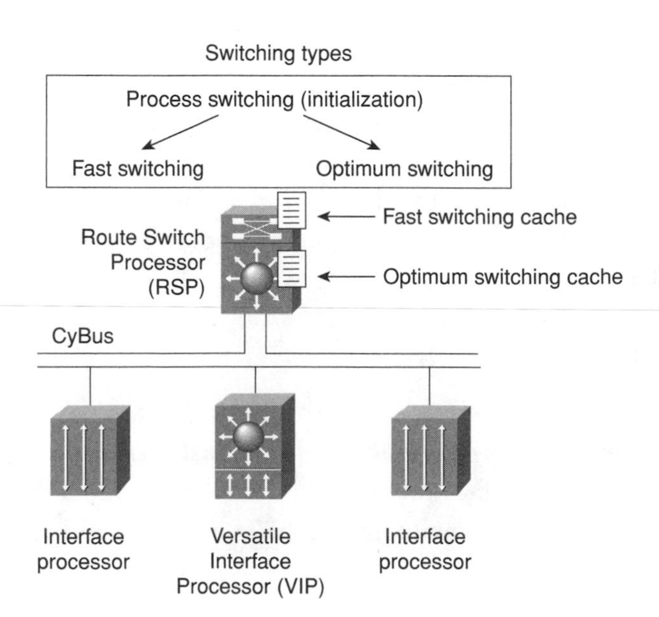

Figure 1–5 Distributed Switching on Cisco 7500 Series Routers

The VIP card installed in this router maintains a copy of the routing cache information needed to forward packets. Because the VIP card has the routing information it needs, it performs the switching locally, making the packet forwarding much faster. Router throughput is increased linearly based on the number of VIP cards installed in the router.

NetFlow Switching

NetFlow switching enables you to collect the data required for flexible and detailed accounting, billing, and chargeback for network and application resource utilization. Accounting data can be collected for both dedicated line and dial-access accounting. NetFlow switching over a foundation of VLAN technologies provides the benefits of switching and routing on the same platforms. NetFlow switching is supported over switched LAN or ATM backbones, allowing scalable inter-VLAN forwarding. NetFlow switching can be deployed at any location in the network as an extension to existing routing infrastructures.

Platform and Switching Path Correlation

Depending on the routing platform you are using, availability and default implementations of switching paths vary. Table 1–1 shows the correlation between Cisco IOS switching paths and routing platforms.

Table 1–1 *Switching Paths on RSP-Based Routers*

Switching Path	Cisco 7200	Cisco 7500	Comments	Configuration Command
Process switching	Yes	Yes	Initializes switching caches	**no** *protocol* **route-cache**
Fast switching	Yes	Yes	Default (except for IP)	*protocol* **route-cache**
Optimum switching	Yes	Yes	Default for IP	*protocol* **route-cache optimum**
Distributed switching	No	Yes	Using Second-Generation VIP line cards	*protocol* **route-cache distributed**
NetFlow switching	Yes	Yes	Configurable per interface	*protocol* **route-cache flow**

Understanding Features that Affect Performance

Performance is derived from the switching mechanism you are using. Some Cisco IOS features require special handling and cannot be switched until the additional processing they require has been performed. This special handling is not processing that the interface processors can perform. Because these features require additional processing, they affect switching performance. These features include the following:

- Queuing
- Random Early Detection
- Compression
- Filtering (using access lists)
- Encryption
- Accounting

Queuing

Queuing occurs when network congestion transpires. When traffic is moving well within the network, packets are sent as they arrive at the interface. Cisco IOS software implements four different queuing algorithms:

- First In, First Out (FIFO) Queuing—Packets are forwarded in the same order in which they arrive at the interface.
- Priority Queuing—Packets are forwarded based on an assigned priority. You can create priority lists and groups to define rules for assigning packets to priority queues.
- Custom Queuing—You can control a percentage of interface bandwidth for specified traffic by creating protocol queue lists and custom queue lists.
- Weighted Fair Queuing—Provides automatic traffic priority management. Low-bandwidth sessions have priority over high-bandwidth sessions, and high-bandwidth sessions are assigned weights. Weighted Fair Queuing is the default for interfaces slower than 2.048Mbps.

Random Early Detection

Random Early Detection is designed for congestion avoidance. Traffic is prioritized based on type of service (TOS) or precedence. This feature is available on T3, OC-3, and ATM interfaces.

Compression

Depending on the protocol you are using, various compression options are available in Cisco IOS software. Refer to the Cisco IOS configuration guide for the protocol you are using to see what compression options you have.

Filtering

You can define access lists to control access to or from a router for a number of services. You could, for example, define an access list to prevent packets with a certain IP address from leaving a particular interface on a router. How access lists are used depends on the protocol.

Encryption

Encryption algorithms are applied to data to alter its appearance, making it incomprehensible to those who are not authorized to see the data.

Accounting

You can configure accounting features to collect network data related to resource usage. The information you collect (in the form of statistics) can be used for billing, chargeback, and planning resource usage.

Configuring Fast Switching

Fast switching allows higher throughput by switching a packet using a cache created by the initial packet that was sent to a particular destination. Destination addresses are stored in the high-speed cache to expedite forwarding. Routers offer better packet-transfer performance when fast switching is enabled. Fast switching is enabled by default on all interfaces that support it.

To configure appropriate fast-switching features, perform the tasks in these sections:

- Enable AppleTalk Fast Switching
- Enable IP Fast Switching
- Enable Fast Switching on the Same IP Interface
- Enable Fast Switching of IPX-Directed Broadcast Packets

NOTES

Fast switching is not supported for the X.25 encapsulations.

Enable AppleTalk Fast Switching

AppleTalk access lists are automatically fast switched. Access list fast switching improves the performance of AppleTalk traffic when access lists are defined on an interface.

Enable IP Fast Switching

Fast switching involves the use of a high-speed switching cache for IP routing. Destination IP addresses are stored in the high-speed cache to expedite packet forwarding. In some cases, fast switching is inappropriate, such as when slow-speed serial links (64KB and lower) are being fed from higher-speed media, such as T1 or Ethernet. In such a case, disabling fast switching can reduce the packet drop rate to some extent. Fast switching allows outgoing packets to be load balanced on a *per-destination* basis.

To enable or disable fast switching, perform either of the following tasks in interface configuration mode:

Task	Command
Enable fast switching (use of a high-speed route cache for IP routing).	**ip route-cache**
Disable fast switching and enable load balancing on a per-packet basis.	**no ip route-cache**

Enable Fast Switching on the Same IP Interface

You can enable IP fast switching when the input and output interfaces are the same interface. This normally is not recommended; however, it is useful when you have partially meshed media, such as Frame Relay. You also could use this feature on other interfaces, although it is not recommended because it would interfere with redirection.

Figure 1–6 illustrates a scenario in which this is desirable. Router A has a data link connection identifier (DLCI) to Router B, and Router B has a DLCI to Router C. There is no DLCI between Routers A and C; traffic between them must go in and out of Router B through the same interface.

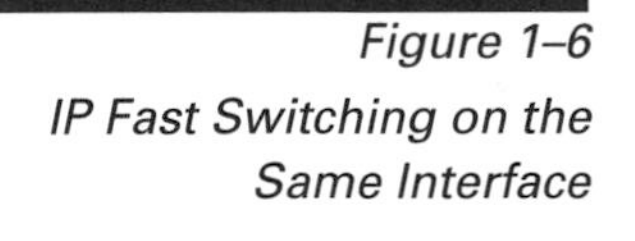

Figure 1–6
IP Fast Switching on the Same Interface

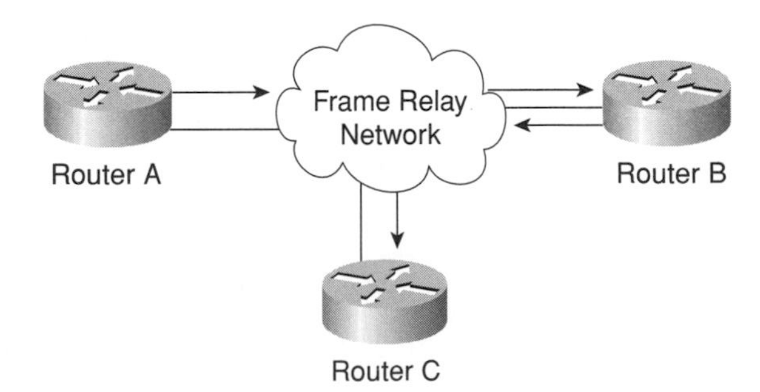

To allow IP fast switching on the same interface, perform the following task in interface configuration mode:

Task	Command
Enable the fast switching of packets out of the same interface on which they arrived.	**ip route-cache same-interface**

Enable Fast Switching of IPX-Directed Broadcast Packets

By default, Cisco IOS software switches packets that have been directed to the broadcast address. To enable fast switching of these IPX-directed broadcast packets, perform the following task in global configuration mode:

Task	Command
Enable fast switching of IPX-directed broadcast packets.	**ipx broadcast-fastswitching**

Enable SMDS Fast Switching

SMDS fast switching of IP, IPX, and AppleTalk packets provides faster packet transfer on serial links with speeds higher than 56 Kbps. Use fast switching if you use high-speed, packet-switched, datagram-based WAN technologies offered by service providers, such as Frame Relay.

By default, SMDS fast switching is enabled. To re-enable fast switching, if it has been disabled, perform the following tasks in interface configuration mode:

Task		Command
Step 1	Define the type and unit number of the interface and enter interface configuration mode.	**interface** *type number*
Step 2	Set SMDS encapsulation.	**encapsulation smds**
Step 3	Enable the interface for IP fast switching.	**ip route-cache**
Step 4	Enable the interface for IPX fast switching.	**ipx route-cache**
Step 5	Enable the interface for AppleTalk fast switching.	**appletalk route-cache**

Disabling Fast Switching for Troubleshooting

Fast switching allows higher throughput by switching packets using a cache created by previous packets. Packet transfer performance generally is better when fast switching is enabled. Fast switching also provides load sharing on a per-packet basis.

By default, fast switching is enabled on all interfaces that support it. You may, however, want to disable fast switching to save memory space on interface cards and to avoid congestion when high-bandwidth interfaces are writing large amounts of information to low-bandwidth interfaces. This is especially important when using rates slower than T1.

NOTES

Fast switching is not supported on serial interfaces using encapsulations other than HDLC.

TIMESAVER

Turning off fast switching increases system overhead.

For some diagnostics, such as debugging and packet-level tracing, you need to disable fast switching. If fast switching is running, you do not see packets unless they pass through the route processor. Otherwise, packets would be switched on the interface. You might want to turn off fast switching temporarily and bypass the route processor while you are trying to capture information.

This section covers the following topics:

- Disable AppleTalk Fast Switching
- Disable Banyan VINES Fast Switching
- Disable DECnet Fast Switching
- Disable IPX Fast Switching
- Disable ISO CLNS Fast Switching Through the Cache
- Disable XNS Fast Switching

Disable AppleTalk Fast Switching

To disable AppleTalk fast switching on an interface, perform the following task in interface configuration mode:

Task	Command
Disable AppleTalk fast switching.	**no appletalk route-cache**

Disable Banyan VINES Fast Switching

By default, fast switching is enabled on all interfaces on which it is supported. To disable fast switching on an interface, perform the following task in interface configuration mode:

Task	Command
Disable fast switching.	**no vines route-cache**

Disable DECnet Fast Switching

By default, Cisco's DECnet routing software implements fast switching of DECnet packets. To disable fast switching of DECnet packets, perform the following task in interface configuration mode:

Task	Command
Disable fast switching of DECnet packets on a per-interface basis.	**no decnet route-cache**

Disable IPX Fast Switching

To disable IPX fast switching, perform the following task in interface configuration mode:

Task	Command
Disable IPX fast switching.	**no ipx route-cache**

Disable ISO CLNS Fast Switching Through the Cache

By default, ISO CLNS fast switching through the cache is enabled for all supported interfaces. To disable fast switching, perform the following task in interface configuration mode:

Task	Command
Disable fast switching.	**no clns route-cache**

NOTES

The cache still exists. It is used after the **no clns route-cache** interface configuration command is used; the software just does not perform fast switching through the cache.

Disable XNS Fast Switching

To disable XNS fast switching on an interface, perform the following task in interface configuration mode:

Task	Command
Disable XNS fast switching.	**no xns route-cache**

DISABLING OPTIMUM SWITCHING FOR TROUBLESHOOTING

By default, optimum switching is enabled for IP on Ethernet, FDDI, serial interfaces on the Cisco 7500 series, and on all ATM port adapter interfaces. On serial interfaces, it is supported for HDLC encapsulation only. To use fast switching or process switching, disable optimum switching by performing the following task in interface configuration mode:

Task	Command
Disable optimum switching.	**no ip route-cache optimum**

CAUTION

Optimum switching must be disabled for troubleshooting.

CONTROLLING THE ROUTE CACHE

The high-speed route cache used by IP fast switching is invalidated when the IP routing table changes. By default, the invalidation of the cache is delayed slightly to avoid excessive CPU load while the routing table is changing. To control the route cache, perform the appropriate tasks in these sections:

- Control Route Cache Invalidation for IP
- Display System and Network Statistics
- Adjust the Route Cache for IPX
- Pad Odd-Length IPX Packets

Control Route Cache Invalidation for IP

To control route cache invalidation, perform the following tasks in global configuration mode as needed for your network:

Task	Command
Allow immediate invalidation of the cache.	**no ip cache-invalidate-delay**
Delay invalidation of the cache.	**ip cache-invalidate-delay** [*minimum maximum quiet threshold*]

CAUTION

The preceding task normally should not be necessary. It should be performed only under the guidance of technical staff. Incorrect configuration can seriously degrade the performance of your router.

Display System and Network Statistics

You can display the contents of IP routing tables and caches. The resulting information can be used to determine resource utilization and to solve network problems.

Perform the following task in privileged EXEC mode:

Task	Command
Display the routing table cache used to fast switch IP traffic.	**show ip cache** [*prefix mask*] [*type number*]

Adjust the Route Cache for IPX

Adjusting the route cache enables you to control the size of the route cache, reduce memory consumption, and improve router performance. You accomplish these tasks by controlling the route cache size and invalidation. The following sections describe these optional tasks:

- Control IPX Route Cache Size
- Control IPX Route Cache Invalidation

Control IPX Route Cache Size

You can limit the number of entries stored in the IPX route cache to free up router memory and aid router processing. Storing too many entries in the route cache can use a significant amount of router memory, causing router processing to slow. This situation is most common on large networks that run network management applications for NetWare.

For example, if a network management station is responsible for managing all clients and servers in a very large (greater than 50,000 nodes) Novell network, the routers on the local segment can become inundated with route cache entries. You can set a maximum number of route cache entries on these routers to free up router memory and aid router processing.

To set a maximum limit on the number of entries in the IPX route cache, complete this task in global configuration mode:

Task	Command
Set a maximum limit on the number of entries in the IPX route cache.	**ipx route-cache max-size** *size*

If the route cache has more entries than the specified limit, the extra entries are not deleted. If route cache invalidation is in use, however, they may be removed. See the following section ("Control IPX Route Cache Invalidation") for more information on invalidating route cache entries.

Control IPX Route Cache Invalidation

You can configure the router to invalidate fast switch cache entries that are inactive. If these entries remain invalidated for one minute, the router purges them from the route cache.

Purging invalidated entries reduces the size of the route cache, reduces memory consumption, and improves router performance. Purging entries also helps ensure accurate route cache information.

You specify the period of time that valid fast switch cache entries must be inactive before the router invalidates them. You can also specify the number of cache entries that the router can invalidate per minute.

To configure the router to invalidate fast switch cache entries that are inactive, complete this task in global configuration mode:

Task	Command
Invalidate fast switch cache entries that are inactive.	**ipx route-cache inactivity-timeout** *period* [*rate*]

NOTES

When you use the **ipx route-cache inactivity-timeout** command with the **ipx route-cache max-size** command, you will have a small route cache with fresh entries.

Pad Odd-Length IPX Packets

Some IPX end hosts accept only even-length Ethernet packets. If the length of a packet is odd, the packet must be padded with an extra byte so that the end host can receive it. By default, Cisco IOS pads odd-length Ethernet packets. There are cases in certain topologies, however, in which non-padded Ethernet packets are being forwarded onto a remote Ethernet network. Under specific conditions, you can enable padding on intermediate media as a temporary workaround for this problem. Note that you should perform this task only under the guidance of a customer engineer or other service representative.

To enable the padding of odd-length packets, perform the following tasks in interface configuration mode:

Task	Command
Step 1 Disable fast switching.	**no ipx route-cache**
Step 2 Enable the padding of odd-length packets.	**ipx pad-process-switched-packets**

PART 2

NetFlow Switching

CHAPTER 2

Configuring NetFlow Switching

This chapter describes NetFlow switching. For documentation of other commands that appear in this chapter, you can use the command reference master index or search online. This chapter contains the following sections:

- Understanding NetFlow Switching
- Configuring NetFlow Switching
- NetFlow Switching Configuration Example

UNDERSTANDING NETFLOW SWITCHING

NetFlow switching provides network administrators with access to "call detail recording" information from their data networks. Exported NetFlow data can be used for a variety of purposes, including network management and planning, enterprise accounting and departmental chargebacks, ISP billing, data warehousing, and data mining for marketing purposes. NetFlow also provides a highly efficient mechanism with which to process security access lists without paying as much of a performance penalty as is incurred with other available switching methods.

This chapter describes NetFlow switching and how to configure NetFlow features. It contains the following sections:

- NetFlow Switching Support
- Accounting Statistics
- NetFlow Data Format
- Configuring NetFlow Switching

NetFlow Switching Support

NetFlow switching is supported on Cisco 7200 series routers Cisco 7500 series routers.

Accounting Statistics

NetFlow switching is a high-performance, network-layer switching path that captures as part of its switching function a rich set of traffic statistics. These traffic statistics include user, protocol, port, and type of service information that can be used for a wide variety of purposes such as network analysis and planning, accounting, and billing.

NetFlow switching is supported on IP and IP-encapsulated traffic over all interface types and encapsulations, except for ISL/VLAN, ATM, Frame Relay interfaces (when more than one input access control list is used on the interface), and ATM LANE.

Capturing Traffic Data

In conventional switching at the network layer, each incoming packet is handled on an individual basis, with a series of functions to perform access list checks, to capture accounting data, and to switch the packet. With NetFlow switching, after a flow has been identified and access list processing of the first packet in the flow has been performed, all subsequent packets are handled on a "connection-oriented" basis as part of the flow, where access list checks are bypassed and packet switching and statistics capture are performed in tandem.

A network flow is identified as a unidirectional stream of packets between a given source and destination—both defined by a network-layer IP address and transport-layer port number. Specifically, a flow is identified as the combination of the following fields:

- Source IP address
- Destination IP address
- Source port number
- Destination port number
- Protocol type
- Type of service
- Input interface

NetFlow Cache

NetFlow switching operates by creating a flow cache that contains the information needed to switch and perform access list check for all active flows. The NetFlow cache is built by processing the first packet of a flow through the standard switching path (fast or optimum). As a result, each flow is associated with an incoming and outgoing interface port number and with a specific security access permission and encryption policy. The cache also includes entries for traffic statistics that are updated in tandem with the switching of subsequent packets. After the NetFlow cache is created, packets identified as belonging to an existing flow can be switched based on the cached information and security access list checks bypassed. Flow information is maintained within the NetFlow cache for all active flows.

NetFlow Data Format

NetFlow exports flow information in UDP datagrams in one of two formats. The version 1 format was the initial released version, and version 5 is a later enhancement to add Border Gateway Protocol (BGP) autonomous system (AS) information and flow sequence numbers. Versions 2 through 4 were not released.

In version 1 and version 5 format, the datagram consists of a header and one or more flow records. The first field of the header contains the version number of the export datagram. Typically, a receiving application that accepts either format allocates a buffer big enough for the largest possible datagram from either format and uses the version from the header to determine how to interpret the datagram. The second field in the header is the number of records in the datagram and should be used to index through the records.

All fields in either version 1 or version 5 formats are in network byte order. Table 2–1 and Table 2–2 describe the data format for version 1, and Table 2–2 and Table 2–3 describe the data format for version 5.

Cisco recommends that receiving applications sanity check datagrams to ensure that the datagrams are from a valid NetFlow source. We recommend you first check the size of the datagram to make sure it is at least long enough to contain the version and count fields. Next, we recommend you verify that the version is valid (1 or 5) and that the number of received bytes is enough for the header and count flow records (using the appropriate version).

Because NetFlow export uses User Datagram Protocol (UDP) to send export datagrams, it is possible for datagrams to be lost. To determine whether flow export information is lost, the version 5 header format contains a flow sequence number. The sequence number is equal to the sequence number of the previous number, plus the number of flows in the previous datagram. After receiving a new datagram, the receiving application can subtract the expected sequence number from the sequence number in the header to get the number of missed flows. Table 2–1 lists the bytes for version 1 header format.

Table 2–1 *Version 1 Header Format*

Bytes	Content	Description
0–3	version and count	Netflow export format version number and number of flows exported in this packet (1–24)
4–7	SysUptime	Current time in milliseconds since router booted
8–11	unix_secs	Current seconds since 0000 UTC 1970
12–16	unix_nsecs	Residual nanoseconds since 0000 UTC 1970

Table 2–2 lists the byte definitions for version 1 flow record format.

Table 2–2 *Version 1 Flow Record Format*

Bytes	Content	Description
0–3	srcaddr	Source IP address
4–7	dstaddr	Destination IP address
8–11	nexthop	Next hop router's IP address
12–15	input and output	Input and output interface's SNMP index
16–19	dPkts	Packets in the flow
20–23	dOctets	Total number of layer 3 bytes in the flow's packets
24–27	First	SysUptime at start of flow
28–31	Last	SysUptime at the time the last packet of flow was received
32–35	srcport and dstport	TCP/UDP source and destination port number or equivalent
36–39	pad1, prot, and tos	Unused (zero) byte, IP protocol (for example, 6=TCP, 17=UDP), and IP type-of-service
40–43	flags, pad2, and pad3	Cumulative OR of TCP flags. Pad 2 and pad 3 are unused (zero) byte
44–48	reserved	Unused (zero) bytes

Table 2–3 lists the byte definitions for version 5 header format.

Table 2–3 *Version 5 Header Format*

Bytes	Content	Description
0–3	version and count	Netflow export format version number and number of flows exported in this packet (1–30)
4–7	SysUptime	Current time in milliseconds since router booted
8–11	unix_secs	Current seconds since 0000 UTC 1970
12–15	unix_nsecs	Residual nanoseconds since 0000 UTC 1970

Table 2–3 *Version 5 Header Format, Continued*

Bytes	Content	Description
16–19	flow_sequence	Sequence counter of total flows seen
20–24	reserved	Unused (zero) bytes

Table 2–4 lists the byte definitions for version 5 flow record format.

Table 2–4 *Version 5 Flow Record Format*

Bytes	Content	Description
0–3	srcaddr	Source IP address
4–7	dstaddr	Destination IP address
8–11	nexthop	Next hop router's IP address
12–15	input and output	Input and output interface's SNMP index
16–19	dPkts	Packets in the flow
20–23	dOctets	Total number of layer 3 bytes in the flow's packets
24–27	First	SysUptime at start of flow
28–31	Last	SysUptime at the time the last packet of flow was received
32–35	srcport and dstport	TCP/UDP source and destination port number or equivalent
36–39	pad1, tcp_flags, prot, and tos	Unused (zero) byte, Cumulative OR of TCP flags, IP protocol (for example, 6=TCP, 17=UDP), and IP type-of service
40–43	src_as and dst_as	As of the source and destination, either origin or peer
44–48	src_mask, dst_mask, and pad2	Source and destination address prefix mask bits, pad 2 is unused (zero) bytes

CONFIGURING NETFLOW SWITCHING

NetFlow switching is one of the four available switching modes. When you configure NetFlow on an interface, the other switching modes are not used on that interface. Optimum switching remains the most efficient switching mode and results in the highest throughput when extensive access list processing is not required. NetFlow comes in a close second (within 15 to 20 percent of optimum switching throughput, possibly higher when access lists are involved). Fast switching is third fastest, with process switching the slowest of all. With NetFlow switching, you also can export data (traffic statistics) to a remote workstation for further processing.

NetFlow switching is based on identifying packet flows and performing switching and access list processing within a router. It does not involve any connection-setup protocol either between routers or to any other networking device or end station and does not require any change externally—either to the traffic or packets themselves or to any other networking device. Thus, NetFlow switching is completely transparent to the existing network, including end stations and application software and network devices such as LAN switches. Also, because NetFlow switching is performed independently on each internetworking device, it does not need to be operational on each router in the network. Network planners can selectively invoke NetFlow switching (and NetFlow data export) on a router or interface basis to gain traffic performance, control, or accounting benefits in specific network locations.

NOTES

NetFlow does consume additional memory and CPU resources compared to other switching modes; therefore, it is important to understand the resources required on your router before enabling NetFlow.

To configure NetFlow switching, first configure the router for IP routing. After you configure IP routing, perform the following tasks beginning in global configuration mode:

Task		Command
Step 1	Specify the interface, and enter interface configuration mode.	**interface** *type slot/port-adapter /port* (Cisco 7500 series routers) **interface** *type slot/port* (Cisco 7200 series routers)
Step 2	Specify flow switching.	**ip route-cache flow**

NetFlow switching information can also be exported to network management applications. To configure the router to export NetFlow switching statistics maintained in the NetFlow cache to a workstation when a flow expires, perform one of the following tasks in global configuration mode:

Task	Command
Configure the router to export NetFlow cache entries to a workstation if you are using receiving software that requires version 1. Version 1 is the default.	**ip flow-export** *ip-address udp-port* [**version 1**]
Configure the router to export NetFlow cache entries to a workstation if you are using receiving software that accepts version 5. Optionally specify origin or peer autonomous system (AS). The default is to export neither AS which provides improved performance.	**ip flow-export** *ip-address udp-port* **version 5** [**origin-as** | **peer-as**]

Normally the size of the NetFlow cache will meet your needs. You can, however, increase or decrease the number of entries maintained in the cache to meet the needs of your NetFlow traffic rates. The default is 64K flow cache entries. Each cache entry is approximately 64 bytes of storage. Assuming a cache with the default number of entries, approximately 4MB of DRAM would be required. Each time a new flow is taken from the free-flow queue, the number of free flows is checked. If only a few free flows remain, NetFlow attempts to age 30 flows using an accelerated timeout. If only one free flow remains, NetFlow automatically ages 30 flows, regardless of their age. The intent is to ensure free flow entries are always available.

To customize the number of entries in the NetFlow cache, perform the following task in global configuration mode:

Task	Command
Change the number of entries maintained in the NetFlow cache. The number of entries can be 1024 to 524288. The default is 65536.	**ip flow-cache entries** *number*

CAUTION

Cisco recommends that you not change the NetFlow cache entries. Improper use of this feature could cause network problems. To return to the default NetFlow cache entries, use the **no ip flow-cache entries** global configuration command.

Manage NetFlow Switching Statistics

You can display and clear NetFlow switching statistics. NetFlow statistics consist of IP packet size distribution, IP flow switching cache information, and flow information such as the protocol, total flow, flows per second, and so forth. The resulting information can be used to find out information about your router traffic. To manage NetFlow switching statistics, perform any of the following tasks in privileged EXEC mode:

Task	Command
Display the NetFlow switching statistics.	**show ip route flow**
Clear the NetFlow switching statistics.	**clear ip flow stats**

Configuring IP Distributed and NetFlow Switching on VIP Interfaces

On Cisco 7500 series routers with a Route Switch Processor (RSP) and with Versatile Interface Processor (VIP) controllers, the VIP hardware can be configured to switch packets received by the VIP with no per-packet intervention on the part of the RSP. This process is called *distributed switching*. Distributed switching decreases the demand on the RSP.

The VIP hardware can also be configured for NetFlow switching. A new high-performance feature that identifies initiation of traffic flow between internet endpoints caches information about the flow, and uses this cache for high-speed switching of subsequent packets within the identified stream.

NOTES

NetFlow switching data can also be exported to network management applications.

To configure distributed switching on the VIP, first configure the router for IP routing as described in this chapter and the various routing protocol chapters, depending on the protocols you use.

After you configure IP routing, perform the following tasks beginning in global configuration mode:

Task		Command
Step 1	Specify the interface, and enter interface configuration mode.	**interface** *type slot/port-adapter /port*
Step 2	Enable VIP distributed switching of IP packets on the interface.	**ip route-cache distributed**
Step 3	Specify either flow or optimum switching.	**ip route-cache** [**flow** | **optimum**]

When extended access lists are configured, flow switching is faster than the default optimum fast-switching on Cisco 7507 and 7513 platforms. When the RSP or VIP is flow switching, it uses a flow cache instead of a destination network cache to switch IP packets. The flow cache uses source and destination network address, protocol, and source and destination port numbers to distinguish entries.

To export NetFlow switching cache entries to a workstation when a flow expires, perform the following task in global configuration mode:

Task	Command
Configure the router to export NetFlow cache entries to a workstation.	**ip flow-export** *ip-address udp-port*

To improve performance, fragmented IP packets are optimum or flow switched (depending which switching method is enabled) rather than process switched by default on Cisco 7500 series routers.

NetFlow Switching Configuration Example

The following example shows how to modify the configuration of serial interface 3/0/0 to enable NetFlow switching and to export the flow statistics for further processing to UDP port 0 on a workstation with the IP address of 1.1.15.1. In this example, existing NetFlow statistics are cleared to ensure accurate information when the **show ip cache flow** command is executed to view a summary of the NetFlow switching statistics. Following is the code:

```
configure terminal
interface serial 3/0/0
 ip route-cache flow
 exit
 ip flow-export 1.1.15.1 0 version 5 peer-as
 exit
 clear ip flow stats
```

CHAPTER 3

Cisco IOS Switching Commands

This chapter documents commands used to configure switching features in Cisco IOS software.

NOTES

In Cisco IOS Release 11.3, all commands supported on the Cisco 7500 series routers are also supported on Cisco 7000 series routers.

CLEAR IP FLOW STATS

To clear the NetFlow switching statistics, use the **clear ip flow stats** EXEC command.

clear ip flow stats

Syntax Description

This command has no arguments or keywords.

Command Mode

EXEC

Usage Guidelines

This command first appeared in Cisco IOS Release 11.1 CA.

The **show ip cache flow** command displays the NetFlow switching statistics. Use the **clear ip flow** command to clear the NetFlow switching statistics.

Example

The following example clears the NetFlow switching statistics on the router:

```
clear ip flow stats
```

Related Commands

You can use the master index or search online to find documentation of related commands.

show ip cache

ENCAPSULATION ISL

Use the **encapsulation isl** subinterface configuration command to enable the Inter-Switch Link (ISL). ISL is a Cisco protocol for interconnecting multiple switches and routers, and for defining VLAN topologies.

encapsulation isl *vlan-identifier*

Syntax	*Description*
vlan-identifier	Virtual LAN identifier. The allowed range is 1 to 1000.

Default

Disabled

Command Mode

Subinterface configuration

Usage Guidelines

This command first appeared in Cisco IOS Release 11.1.

ISL encapsulation is configurable on Fast Ethernet interfaces.

ISL encapsulation adds a 26-byte header to the beginning of the Ethernet frame. The header contains a 10-bit VLAN identifier that conveys VLAN membership identities between switches.

Example

The following example enables ISL on Fast Ethernet subinterface 2/1.20:

```
interface FastEthernet 2/1.20
 encapsulation isl 400
```

Related Commands

You can use the master index or search online to find documentation of related commands.

bridge-group
debug vlan
show bridge vlan
show interfaces
show vlans

ENCAPSULATION SDE

Use the **encapsulation sde** subinterface configuration command to enable IEEE 802.10 encapsulation of traffic on a specified subinterface in virtual LANs. IEEE 802.10 is a standard protocol for interconnecting multiple switches and routers and for defining VLAN topologies.

encapsulation sde *said*

Syntax	*Description*
said	Security association identifier. This value is used as the virtual LAN identifier. The valid range is 0 through 0xFFFFFFFE.

Default

Disabled

Command Mode

Subinterface configuration

Usage Guidelines

This command first appeared in Cisco IOS Release 10.3.

SDE encapsulation is configurable only on the following interface types:

IEEE 802.10 Routing	IEEE 802.10 Transparent Bridging
• FDDI	• Ethernet • FDDI • HDLC Serial • Transparent mode • Token Ring

Example

The following example enables SDE on FDDI subinterface 2/0.1 and assigns a VLAN identifier of 9999:

```
interface fddi 2/0.1
 encapsulation sde 9999
```

Related Commands

You can use the master index or search online to find documentation of related commands.

bridge-group
debug vlans
show bridge vlan
show interfaces
show vlans

IP CACHE-INVALIDATE-DELAY

To control the invalidation rate of the IP route cache, use the **ip cache-invalidate-delay** global configuration command. To allow the IP route cache to be immediately invalidated, use the **no** form of this command.

ip cache-invalidate-delay [*minimum maximum quiet threshold*]
no ip cache-invalidate-delay

Syntax	*Description*
minimum	(Optional) Minimum time (in seconds) between invalidation request and actual invalidation. The default is 2 seconds.
maximum	(Optional) Maximum time (in seconds) between invalidation request and actual invalidation. The default is 5 seconds.
quiet	(Optional) Length of quiet period (in seconds) before invalidation.
threshold	(Optional) Maximum number of invalidation requests considered to be quiet.

Defaults

minimum = 2 seconds
maximum = 5 seconds, and 3 seconds with no more than zero invalidation requests

Command Mode

Global configuration

Usage Guidelines

This command first appeared in Cisco IOS Release 10.0.

All cache invalidation requests are honored immediately.

This command should typically not be used except under the guidance of technical support personnel. Incorrect settings can seriously degrade network performance.

The IP fast-switching and autonomous-switching features maintain a cache of IP routes for rapid access. When a packet is to be forwarded and the corresponding route is not present in the cache, the packet is process-switched and a new cache entry is built. However, when routing table changes occur (such as when a link or an interface goes down), the route cache must be flushed so that it can be rebuilt with up-to-date routing information.

This command controls how the route cache is flushed. The intent is to delay invalidation of the cache until after routing has settled down. Because route table changes tend to be clustered in a short period of time, and the cache may be flushed repeatedly, a high CPU load might be placed on the router.

When this feature is enabled, and the system requests that the route cache be flushed, the request is held for at least *minimum* seconds. Then the system determines whether the cache has been "quiet" (that is, less than *threshold* invalidation requests in the last *quiet* seconds). If the cache has been quiet, the cache is then flushed. If the cache does not become quiet within *maximum* seconds after the first request, it is flushed unconditionally.

Manipulation of these parameters trades off CPU utilization versus route convergence time. Timing of the routing protocols is not affected, but removal of stale cache entries is affected.

Example

The following example sets a minimum delay of 5 seconds, a maximum delay of 30 seconds, and a quiet threshold of no more than 5 invalidation requests in the previous 10 seconds:

```
ip cache-invalidate-delay 5 30 10 5
```

Related Commands

You can use the master index or search online to find documentation of related commands.

ip route-cache
show ip cache

IP FLOW-CACHE ENTRIES

Use the **ip flow-cache entries** global configuration command to change the number of entries maintained in the NetFlow cache. Use the **no** form of this command to return to the default number of entries.

ip flow-cache entries *number*
no ip flow-cache entries

Syntax	*Description*
number	Number of entries to maintain in the NetFlow cache. Range is 1024 to 524288 entries. The default is 65536 (64K).

Default

65536 entries (64K)

Command Mode

Global configuration

Usage Guidelines

This command first appeared in Cisco IOS Release 11.1 CA.

Normally, the default size of the NetFlow cache will meet your needs; however, you can increase or decrease the number of entries maintained in the cache to meet the needs of your flow traffic rates. For environments with a high amount of flow traffic (such as an internet core router), a larger value, such as 131072 (128K), is recommended. To obtain information on your flow traffic, use the **show ip cache flow** command.

The default is 64K flow cache entries. Each cache entry is approximately 64 bytes of storage. Assume that a cache with the default number of entries, approximately 4MB of DRAM, would be required. Each time a new flow is taken from the free flow queue, the number of free flows is checked. If there are only a few free flows remaining, NetFlow attempts to age 30 flows using an accelerated timeout. If there is only one free flow remaining, NetFlow automatically ages 30 flows, regardless of their age. The intent is to ensure free flow entries are always available.

CAUTION

Cisco recommends that you do not change the NetFlow cache entries. Improper use of this feature could cause network problems. To return to the default NetFlow cache entries, use the **no ip flow-cache entries** global configuration command.

Example

The following example increases the number of entries in the NetFlow cache to 131072 (128K):

```
ip flow-cache entries 131072
```

Related Commands

You can use the master index or search online to find documentation of related commands.

show ip cache

IP FLOW-EXPORT

To enable the exporting of information in NetFlow cache entries, use the **ip flow-export** global configuration command. To disable the exporting of information, use the **no** form of this command.

ip flow-export *ip-address udp-port* [**version 1** | **version 5** [**origin-as** | **peer-as**]]
no ip flow-export

Syntax	*Description*
ip-address	IP address of the workstation to which you want to send the NetFlow information.
udp-port	UDP protocol-specific port number.
version 1	(Optional) Specifies that the export packet uses the version 1 format. This is the default. The version field occupies the first two bytes of the export record. The number of records stored in the datagram is a variable between 1 and 24 for version 1.
version 5	(Optional) Specifies export packet uses the version 5 format. The number of records stored in the datagram is a variable between 1 and 30 for version 5.
origin-as	(Optional) Specifies that export statistics includes the origin autonomous system (AS) for the source and destination.
peer-as	(Optional) Specifies that export statistics includes the peer AS for the source and destination.

Default

Disabled

Command Mode

Global configuration

Usage Guidelines

This command first appeared in Cisco IOS Release 11.1.

This command was modified to include the **version** keyword in Cisco IOS Release 11.1 CA.

There is a lot of information in a NetFlow cache entry. When flow switching is enabled with the **ip route-cache flow** command, you can use the **ip flow-export** command to configure the router to export the flow cache entry to a workstation when a flow expires. This feature can be useful for purposes of statistics, billing, and security.

Version 5 format includes the source and destination AS addresses, source and destination prefix masks, and a sequence number. Because this change may appear on your router as a maintenance release, support for version 1 format is maintained with the **version 1** keyword.

Examples

The following example configures the router to export the NetFlow cache entry to UDP port 125 on the workstation at 134.22.23.7 when the flow expires using version 1 format:

```
ip flow-export 134.22.23.7 125
```

The following example configures the router to export the NetFlow cache entry to UDP port 2048 on the workstation at 134.22.23.7 when the flow expires using version 5 format and including the peer AS information:

```
ip flow-export 134.22.23.7 2048 version 5 peer-as
```

Related Command

You can use the master index or search online to find documentation of related commands.

ip route-cache flow

IP ROUTE-CACHE

Use the **ip route-cache** interface configuration command to control the use of high-speed switching caches for IP routing. To disable any of these switching modes, use the **no** form of this command.

ip route-cache [**cbus**]
no ip route-cache [**cbus**]
ip route-cache same-interface
no ip route-cache same-interface
ip route-cache [**optimum** | **flow**]
no ip route-cache [**optimum** | **flow**]
ip route-cache distributed
no ip route-cache distributed

Syntax	*Description*
cbus	(Optional) Enables both autonomous switching and fast switching.
same-interface	Enables fast-switching packets to back out of the interface on which they arrived.
optimum	(Optional) Enables optimum fast switching on the Cisco 7500 series. This feature is enabled by default for IP on all supported interfaces (Ethernet, FDDI, and serial). For serial interfaces, it is supported for HDLC encapsulation only.
flow	(Optional) Enables the RSP to perform flow switching on the interface.
distributed	Enables VIP distributed switching on the interface. This feature can be enabled on Cisco 7500 series routers with an RSP and Versatile Interface Processor (VIP) controllers. If both **ip route-cache flow** and **ip route-cache distributed** are configured, the VIP does distributed flow switching. If only **ip route-cache distributed** is configured, the VIP does distributed optimum switching.

Defaults

IP autonomous switching is disabled.
Fast switching varies by interface and media.
Optimum switching is enabled on supported interfaces.
Distributed switching is disabled.

Command Mode

Interface configuration

Usage Guidelines

This command first appeared in Cisco IOS Release 10.0. The **optimum** keyword first appeared in Cisco IOS Release 11.1. The **distributed** keyword first appeared in Cisco IOS Release 11.2.

Using the route cache is often called *fast switching*. The route cache allows outgoing packets to be load-balanced on a *per-destination* basis.

The **ip route-cache** command with no additional keywords enables fast switching.

Our routers generally offer better packet transfer performance when fast switching is enabled, with one exception. On networks using slow serial links (64K and below), disabling fast switching to enable the per-packet load sharing is usually the best choice.

You can enable IP fast switching when the input and output interfaces are the same interface using the **ip route-cache same-interface** command. This normally is not recommended, though it is useful when you have partially meshed media, such as Frame Relay. You could use this feature on other interfaces, although it is not recommended because it will interfere with redirection.

When IP accounting or extended access lists are used, flow switching is faster than the default optimum fast-switching on Cisco 7507 and 7513 platforms. When the Route Switch Processor (RSP) is flow switching, it uses a flow cache instead of a destination network cache to switch IP packets. The flow cache uses source and destination network address, protocol, and source and destination port numbers to distinguish entries.

The flow caching option can also be used to allow statistics to be gathered with a finer granularity. The statistics include IP subprotocols, well-known ports, total flows, average number of packets per flow, and average flow lifetime.

On Cisco 7500 series routers with RSP and Versatile Interface Processor (VIP) controllers, the VIP hardware can be configured to switch packets received by the VIP with no per-packet intervention on the part of the RSP. When VIP distributed switching is enabled, the input VIP interface tries to switch IP packets instead of forwarding them to the RSP for switching. Distributed switching helps decrease the demand on the RSP.

NOTES

Not all switching methods are available on all platforms.

Examples

The following example enables both fast switching and autonomous switching:

```
ip route-cache cbus
```

The following example disables both fast switching and autonomous switching:

```
no ip route-cache
```

The following example turns off autonomous switching only:

```
no ip route-cache cbus
```

The following example enables VIP distributed flow switching on the interface:

```
interface ethernet 0/5/0
 ip address 17.252.245.2 255.255.255.0
 ip route-cache distributed
 ip route-cache flow
```

The following example returns the system to its defaults (fast switching enabled; autonomous switching disabled):

```
ip route-cache
```

Related Commands

You can use the master index or search online to find documentation of related commands.

ip cache-invalidate-delay
show ip cache

IP ROUTE-CACHE FLOW

To enable NetFlow switching for IP routing, use the **ip route-cache flow** interface configuration command. To disable NetFlow switching, use the **no** form of this command.

ip route-cache flow
no ip route-cache flow

Syntax Description

This command has no arguments or keywords.

Defaults

Disabled

Command Mode

Interface configuration

Usage Guidelines

This command first appeared in Cisco IOS Release 11.1.

NetFlow switching is a high-performance, network-layer switching path that captures as part of its switching function a rich set of traffic statistics. These traffic statistics include user, protocol, port, and type-of-service information that can be used for a wide variety of purposes such as network analysis and planning, accounting, and billing. To export NetFlow data, use the **ip flow-export** global configuration command.

NetFlow switching is supported on IP and IP encapsulated traffic over all interface types and encapsulations, except for ISL/VLAN, ATM, and Frame Relay interfaces (when more than one input access control list is used on the interface) and ATM LANE.

In conventional switching at the network layer, each incoming packet is handled on an individual basis with a series of functions to perform access list checks, capture accounting data, and switch the packet. With NetFlow switching, after a flow has been identified and access list processing of the first packet in the flow has been performed, all subsequent packets are handled on a "connection-oriented" basis as part of the flow, where access list checks are bypassed and packet switching and statistics capture are performed in tandem.

A network flow is identified as a unidirectional stream of packets between a source and destination—both defined by a network-layer IP address and transport-layer port number. Specifically, a flow is identified as the combination of the following fields:

- source IP address
- destination IP address

- source port number
- destination port number
- protocol type
- type of service
- input interface

NetFlow switching operates by creating a flow cache that contains the information needed to switch and perform access list check for all active flows. The NetFlow cache is built by processing the first packet of a flow through the standard switching path (fast or optimum). As a result, each flow is associated with an incoming and outgoing interface port number and with a specific security access permission and encryption policy. The cache also includes entries for traffic statistics that are updated in tandem with the switching of subsequent packets. After the NetFlow cache is created, packets identified as belonging to an existing flow can be switched based on the cached information and security access list checks bypassed. Flow information is maintained within the NetFlow cache for all active flows.

NetFlow switching is one of the four available switching modes. When you configure NetFlow on an interface, the other switching modes are not used on that interface. Optimum switching remains the most efficient switching mode and results in the highest throughput when extensive access list processing is not required. NetFlow comes in a close second (within 15 to 2 percent of optimum switching throughput, possibly higher when access lists are involved). Fast switching is third fastest, with process switching the slowest of all. Also, with NetFlow switching you can export data (traffic statistics) to a remote workstation for further processing.

NetFlow switching is based on identifying packet flows and performing switching and access list processing within a router. It does not involve any connection-setup protocol, either between routers or to any other networking device or end station, and does not require any change externally—either to the traffic or packets themselves or to any other networking device. Thus, NetFlow switching is completely transparent to the existing network, including end stations and application software and network devices, such as LAN switches. Also, because NetFlow switching is performed independently on each internetworking device, it does not need to be operational on each router in the network. Network planners can selectively invoke NetFlow switching (and NetFlow data export) on a router/interface basis to gain traffic performance, control, or accounting benefits in specific network locations.

NOTES

NetFlow does consume additional memory and CPU resources compared to other switching modes; therefore, it is important to understand the resources required on your router before enabling NetFlow.

Examples

The following example enables NetFlow switching on the interface:

```
interface ethernet 0/5/0
 ip address 17.252.245.2 255.255.255.0
 ip route-cache flow
```

The following example returns the interface to its defaults (fast switching enabled; autonomous switching disabled):

```
interface ethernet 0/5/0
 ip route-cache
```

Related Commands

You can use the master index or search online to find documentation of related commands.

ip flow-export
show ip cache

SHOW IP CACHE

To display the routing table cache used to fast switch IP traffic, use the **show ip cache** EXEC command.

show ip cache [*prefix mask*] [*type number*]

Syntax	*Description*
prefix	(Optional) Display only the entries in the cache that match the prefix and mask combination.
mask	(Optional) Display only the entries in the cache that match the prefix and mask combination.
type	(Optional) Display only the entries in the cache that match the interface type and number combination.
number	(Optional) Display only the entries in the cache that match the interface type and number combination.

Command Mode

EXEC

Usage Guidelines

This command first appeared in Cisco IOS Release 10.0. The arguments *prefix*, *mask*, *type*, and *number* first appeared in Cisco IOS Release 10.0. The **show ip cache** display shows MAC headers up to 92 bytes.

Sample Displays

The following is sample output from the **show ip cache** command:

```
Router# show ip cache

IP routing cache version 4490, 141 entries, 20772 bytes, 0 hash overflows
Minimum invalidation interval 2 seconds, maximum interval 5 seconds,
   quiet interval 3 seconds, threshold 0 requests
Invalidation rate 0 in last 7 seconds, 0 in last 3 seconds
Last full cache invalidation occurred 0:06:31 ago

Prefix/Length       Age       Interface       MAC Header
131.108.1.1/32      0:01:09   Ethernet0/0     AA000400013400000C0357430800
131.108.1.7/32      0:04:32   Ethernet0/0     00000C01281200000C0357430800
131.108.1.12/32     0:02:53   Ethernet0/0     00000C029FD000000C0357430800
131.108.2.13/32     0:06:22   Fddi2/0         00000C05A3E000000C035753AAAA0300
                                              00000800
131.108.2.160/32    0:06:12   Fddi2/0         00000C05A3E000000C035753AAAA0300
                                              00000800
131.108.3.0/24      0:00:21   Ethernet1/2     00000C026BC600000C03574D0800
131.108.4.0/24      0:02:00   Ethernet1/2     00000C026BC600000C03574D0800
131.108.5.0/24      0:00:00   Ethernet1/2     00000C04520800000C03574D0800
131.108.10.15/32    0:05:17   Ethernet0/2     00000C025FF500000C0357450800
131.108.11.7/32     0:04:08   Ethernet1/2     00000C010E3A00000C03574D0800
131.108.11.12/32    0:05:10   Ethernet0/0     00000C01281200000C0357430800
131.108.11.57/32    0:06:29   Ethernet0/0     00000C01281200000C0357430800
```

Table 3–1 describes significant fields shown in the display.

Table 3–1 *Show IP Cache Field Descriptions*

Field	Description
IP routing cache version	Version number of this table. This number is incremented any time the table is flushed.
Entries	Number of valid entries.
Bytes	Number of bytes of processor memory for valid entries.
Hash overflows	Number of times autonomous switching cache overflowed.
Minimum invalidation interval	Minimum time delay between cache invalidation request and actual invalidation.
Maximum interval	Maximum time delay between cache invalidation request and actual invalidation.
Quiet interval	Length of time between cache flush requests before the cache will be flushed.
Threshold *n* requests	Maximum number of requests that can occur while the cache is considered quiet.
Invalidation rate *n* in last *m* seconds	Number of cache invalidations during the last *m* seconds.
0 in last 3 seconds	Number of cache invalidation requests during the last quiet interval.
Last full cache invalidation occurred *hh:mm:ss* ago	Time since last full cache invalidation was performed.
Prefix/Length	Network reachability information for cache entry.
Age	Age of cache entry.
Interface	Output interface type and number.
MAC Header	Layer 2 encapsulation information for cache entry.

The following is sample output from the **show ip cache** command with a prefix and mask specified:

```
Router# show ip cache 131.108.5.0 255.255.255.0

IP routing cache version 4490, 119 entries, 17464 bytes, 0 hash overflows
Minimum invalidation interval 2 seconds, maximum interval 5 seconds,
   quiet interval 3 seconds, threshold 0 requests
Invalidation rate 0 in last second, 0 in last 3 seconds
Last full cache invalidation occurred 0:11:56 ago

Prefix/Length       Age       Interface       MAC Header
131.108.5.0/24      0:00:34   Ethernet1/2     00000C04520800000C03574D0800
```

The following is sample output from the **show ip cache** command with an interface specified:

```
Router# show ip cache e0/2

IP routing cache version 4490, 141 entries, 20772 bytes, 0 hash overflows
Minimum invalidation interval 2 seconds, maximum interval 5 seconds,
   quiet interval 3 seconds, threshold 0 requests
Invalidation rate 0 in last second, 0 in last 3 seconds
Last full cache invalidation occurred 0:06:31 ago

Prefix/Length       Age       Interface       MAC Header
131.108.10.15/32    0:05:17   Ethernet0/2     00000C025FF500000C0357450800
```

SHOW IP CACHE FLOW

To display a summary of the NetFlow switching statistics, use the **show ip cache flow** EXEC command.

show ip cache flow

Syntax Description

This command has no arguments or keywords.

Command Mode

EXEC

Usage Guidelines

This command first appeared in Cisco IOS Release 11.1.

This command was modified to update the display with the latest information in Cisco IOS Release 11.1 CA.

Sample Display

The following is a sample output from the **show ip cache flow** command.

```
Router# show ip cache flow
IP packet size distribution (12718M total packets):
   1-32   64   96  128  160  192  224  256  288  320  352  384  416  448  480
   .000 .554 .042 .017 .015 .009 .009 .009 .013 .030 .006 .007 .005 .004 .004

    512  544  576 1024 1536 2048 2560 3072 3584 4096 4608
   .003 .007 .139 .019 .098 .000 .000 .000 .000 .000 .000

IP Flow Switching Cache, 4456448 bytes
  65509 active, 27 inactive, 820628747 added
  955454490 ager polls, 0 flow alloc failures
  Exporting flows to 1.1.15.1 (2057)
  820563238 flows exported in 34485239 udp datagrams, 0 failed
  last clearing of statistics 00:00:03

Protocol         Total  Flows   Packets Bytes  Packets Active(Sec) Idle(Sec)
--------         Flows   /Sec     /Flow  /Pkt     /Sec     /Flow     /Flow
TCP-Telnet     2656855    4.3        86    78    372.3      49.6      27.6
TCP-FTP        5900082    9.5         9    71     86.8      11.4      33.1
TCP-FTPD       3200453    5.1       193   461   1006.3      45.8      33.4
TCP-WWW      546778274  887.3        12   325  11170.8       8.0      32.3
TCP-SMTP      25536863   41.4        21   283    876.5      10.9      31.3
TCP-X           116391    0.1       231   269     43.8      68.2      27.3
TCP-BGP          24520    0.0        28   216      1.1      26.2      39.0
TCP-Frag         56847    0.0        24   952      2.2      13.1      33.2
TCP-other     49148540   79.7        47   338   3752.6      30.7      32.2
UDP-DNS      117240379  190.2         3   112    570.8       7.5      34.7
UDP-NTP        9378269   15.2         1    76     16.2       2.2      38.7
UDP-TFTP          8077    0.0         3    62      0.0       9.7      33.2
UDP-Frag         51161    0.0        14   322      1.2      11.0      39.4
UDP-other     45502422   73.8        30   174   2272.7       8.5      37.8
ICMP          14837957   24.0         5   224    125.8      12.1      34.3
IGMP             40916    0.0       170   207     11.3     197.3      13.5
IPINIP            3988    0.0     48713   393    315.2     644.2      19.6
GRE               3838    0.0        79   101      0.4      47.3      25.9
IP-other         77406    0.1        47   259      5.9      52.4      27.0
Total:       820563238 1331.7        15   304  20633.0       9.8      33.0
SrcIf     SrcIPaddress     DstIf     DstIPaddress       Pr SrcP DstP Pkts B/Pk Active
Fd0/0     80.0.0.3         Hs1/0     200.1.9.1          06 0621 0052    7   87    5.9
Fd0/0     80.0.0.3         Hs1/0     200.1.8.1          06 0620 0052    7   87    1.8
Hs1/0     200.0.0.3        Fd0/0     80.1.10.1          06 0052 0621    6   58    1.8
Hs1/0     200.0.0.3        Fd0/0     80.1.1.1           06 0052 0620    5   62    5.9
Fd0/0     80.0.0.3         Hs1/0     200.1.3.1          06 0723 0052   16   68    0.3
HS1/0     200.0.0.3        Fd0/0     80.1.2.1           06 0052 0726    6   58   11.8
Fd0/0     80.0.0.3         Hs1/0     200.1.5.1          06 0726 0052    6   96    0.3
Hs1/0     200.0.0.3        Fd0/0     80.1.4.1           06 0052 0442    3   76    0.3
Hs1/0     200.0.0.3        Fd0/0     80.1.7.1           06 0052 D381   11 1171    0.6
```

Table 3–2 describes the fields in the packet size distribution lines of the output.

Table 3–2 *Packet Size Distribution Field Descriptions*

Field	Description
IP packet size distribution	The two lines below this banner show the percentage distribution of packets by size range. In this display, 55.4 percent of the packets fall in the size range 33 to 64 bytes.

Table 3–3 describes the fields in the flow switching cache lines of the output.

Table 3–3 *Flow Switching Cache Display Field Descriptions*

Field	Description
Bytes	Number of bytes of memory the NetFlow cache uses.
Active	Number of active flows in the NetFlow cache at the time this command was entered.
Inactive	Number of flow buffers allocated in the NetFlow cache, but are not currently assigned to a specific flow at the time this command was entered.
Added	Number of flows created since the start of the summary period.
Ager polls	Number of times the NetFlow code looked at the cache to expire entries (used by Cisco for diagnostics only).
Flow alloc failures	Number of times the NetFlow code tried to allocate a flow but could not.
Exporting flows	IP address and UDP port number of the workstation to which flows are exported.
Flows exported in udp datagrams	Total number of flows exported and the total number of UDP datagrams used to export the flows to the workstation.
Failed	Number of flows that could not be exported by the router because of output interface limitations.
Last clearing of statistics	Standard time output (hh:mm:ss) since the clear ip flow stats command was executed. This time output changes to hours and days after the time exceeds 24 hours.

Table 3–4 describes the fields in the activity-by-protocol lines of the output.

Table 3–4 *Activity-By-Protocol Display Field Descriptions*

Field	Description
Protocol	IP protocol and the "well known" port number as described in RFC 1340.
Total Flows	Number of flows for this protocol since the last time statistics were cleared.
Flows/Sec	Average number of flows for this protocol seen per second; equal to total flows/number of seconds for this summary period.
Packets/Flow	Average number of packets observed for the flows seen for this protocol. Equal to Total Packets for this protocol or number of flows for this protocol for this summary period.
Bytes/Pkt	Average number of bytes observed for the packets seen for this protocol (total bytes for this protocol or the total number of packets for this protocol for this summary period).
Packets/Sec	Average number of packets for this protocol per second (total packets for this protocol) or the total number of seconds for this summary period.
Active(Sec)/Flow	Sum of all the seconds from the first packet to the last packet of an expired flow (for example, TCP FIN, time-out, and so forth) in seconds or total flows for this protocol for this summary period.
Idle(Sec)/Flow	Sum of all the seconds from the last packet seen in each nonexpired flow for this protocol until the time this command was entered, in seconds or total flows for this protocol for this summary period.

Table 3–5 describes the fields in the current flow lines of the output.

Table 3–5 *Current Flow Display Field Descriptions*

Field	Description
SrcIf	Internal port name for the source interface.
SrcIPaddress	Source IP address for this flow.
DstIf	Router's internal port name for the destination interface.
DstIPaddress	Destination IP address for this flow.
Pr	IP protocol; for example, 6=TCP, 17=UDP, ... as defined in RFC 1340.

Table 3–5 *Current Flow Display Field Descriptions, Continued*

Field	Description
SrcP	Source port address, TCP/UDP "well known" port number, as defined in RFC 1340.
DstP	Destination port address, TCP/UDP "well known" port number, as defined in RFC 1340.
Pkts	Number of packets observed for this flow.
B/Pkt	Average observed number of bytes per packet for this flow.
Active	Number of seconds between first and last packet of a flow.

Related Commands

You can use the master index or search online to find documentation of related commands.

ip route-cache
clear ip flow stats

PART 3

Virtual LANs

CHAPTER 4

Overview of Routing Between Virtual LANs

This chapter provides an overview of virtual LANs (VLANs). It describes the encapsulation protocols used for routing between VLANs and provides some basic information about designing VLANs, including the following:

- What Is a VLAN?
- VLAN Colors
- Why Implement VLANs?
- Communicating Between VLANs
- VLAN Interoperability
- Designing Switched VLANs

WHAT IS A VIRTUAL LAN?

A *VLAN* is a switched network that is logically segmented on an organizational basis by functions, project teams, or applications rather than on a physical or geographical basis. For example, all workstations and servers used by a particular workgroup team can be connected to the same VLAN, regardless of their physical connections to the network or the fact that they might be intermingled with other teams. Reconfiguration of the network can be done through software rather than by physically unplugging and moving devices or wires.

A VLAN can be thought of as a broadcast domain that exists within a defined set of switches. A VLAN consists of a number of end systems, either hosts or network equipment (such as bridges and routers), connected by a single bridging domain. The bridging domain is supported on various pieces of network equipment; for example, LAN switches that operate bridging protocols between them with a separate bridge group for each VLAN.

VLANs are created to provide the segmentation services traditionally provided by routers in LAN configurations. VLANs address scalability, security, and network management. Routers in VLAN topologies provide broadcast filtering, security, address summarization, and traffic flow management. None of the switches within the defined group will bridge any frames, not even broadcast frames, between two VLANs. The following key issues need to be considered when designing and building switched LAN internetworks:

- LAN Segmentation
- Security
- Broadcast Control
- Performance
- Network Management
- Communication Between VLANs

LAN Segmentation

VLANs allow logical network topologies to overlay the physical switched infrastructure such that any arbitrary collection of LAN ports can be combined into an autonomous user group or community of interest. The technology logically segments the network into separate Layer 2 broadcast domains whereby packets are switched between ports designated to be within the same VLAN. By containing traffic originating on a particular LAN only to other LANs in the same VLAN, switched virtual networks avoid wasting bandwidth, a drawback inherent to traditional bridged and switched networks in which packets are often forwarded to LANs with no need for them. Implementation of VLANs also improves scalability, particularly in LAN environments that support broadcast- or multicast-intensive protocols and applications that flood packets throughout the network.

Figure 4–1 illustrates the difference between traditional physical LAN segmentation and logical VLAN segmentation.

Figure 4–1
LAN Segmentation and VLAN Segmentation

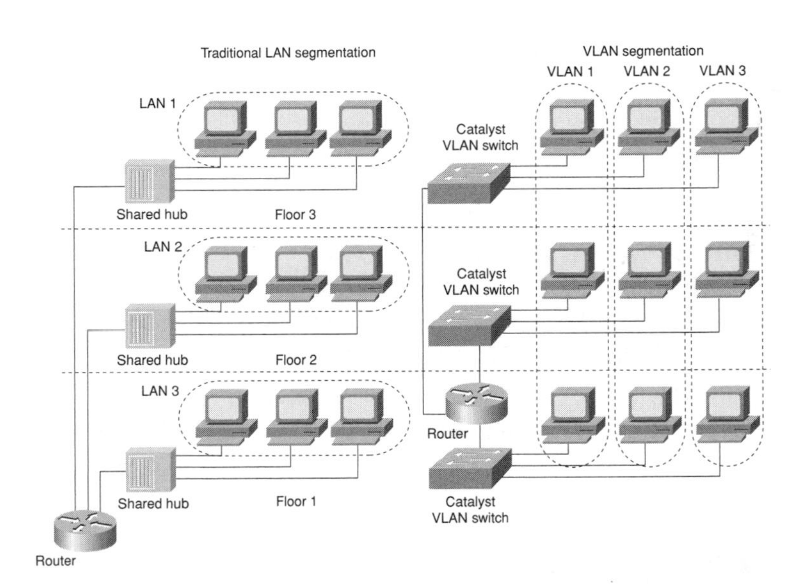

Security

VLANs also improve security by isolating groups. High-security users can be grouped into a VLAN, possibly on the same physical segment, and no users outside that VLAN can communicate with them.

Broadcast Control

Just as switches isolate collision domains for attached hosts and only forward appropriate traffic out a particular port, VLANs provide complete isolation between VLANs. A VLAN is a bridging domain and all broadcast and multicast traffic is contained within it.

Performance

The logical grouping of users allows an accounting group to make intensive use of a networked accounting system assigned to a VLAN that contains just that accounting group and its servers. That group's work will not affect other users. The VLAN configuration improves general network performance by not slowing down other users sharing the network.

Network Management

The logical grouping of users allows easier network management. It is not necessary to pull cables to move a user from one network to another. Adds, moves, and changes are achieved by configuring a port into the appropriate VLAN.

Communication Between VLANs

Communication between VLANs is accomplished through routing, and the traditional security and filtering functions of the router can be used. Cisco IOS software provides network services, such as security filtering, quality of service (QoS), and accounting, on a per VLAN basis. As switched networks evolve to distributed VLANs, Cisco IOS provides key inter-VLAN communications and allows the network to scale.

VLAN Colors

VLAN switching is accomplished through *frame tagging,* where traffic originating and contained within a particular virtual topology carries a unique VLAN identifier (VLAN ID) as it traverses a common backbone or trunk link. The VLAN ID enables VLAN switching devices to make intelligent forwarding decisions based on the embedded VLAN ID. Each VLAN is differentiated by a *color,* or VLAN identifier. The unique VLAN ID determines the *frame coloring* for the VLAN. Packets originating and contained within a particular VLAN carry the identifier that uniquely defines that VLAN (by the VLAN ID).

The VLAN ID allows VLAN switches and routers to selectively forward packets to ports with the same VLAN ID. The switch that receives the frame from the source station inserts the VLAN ID and the packet is switched onto the shared backbone network. When the frame exits the switched LAN, a switch strips the header and forwards the frame to interfaces that match the VLAN color. If you are using a Cisco network management product such as VlanDirector, you can actually color code the VLANs and monitor the VLAN graphically.

Why Implement VLANs?

Network managers can logically group networks that span all major topologies, including high-speed technologies such as ATM, FDDI, and Fast Ethernet. By creating virtual LANs, system and network administrators can control traffic patterns, react quickly to relocations, and keep up with constant changes in the network due to moving requirements and node relocation just by changing the VLAN member list in the router configuration. They can add, remove, or move devices or make other changes to network configuration using software to make the changes.

Issues regarding benefits of creating VLANs should have been addressed when you developed your network design. Issues to consider include the following:

- Scalability
- Performance improvements
- Security
- Network additions, moves, and changes

Communicating Between VLANs

Cisco IOS provides full-feature routing at Layer 3 and translation at Layer 2 between VLANs. Following are the three different protocols available for routing between VLANs:

- Inter-Switch Link (ISL)
- IEEE 802.10
- ATM LAN Emulation

All three of these technologies are based on OSI Layer 2 bridge multiplexing mechanisms.

Inter-Switch Link Protocol

Inter-Switch Link (ISL) protocol is used to inter-connect two VLAN-capable Fast Ethernet devices, such as the Catalyst 5000 or 3000 switches and Cisco 7500 routers. The ISL protocol is a packet-tagging protocol that contains a standard Ethernet frame and the VLAN information associated with that frame. The packets on the ISL link contain a standard Ethernet, FDDI, or token-ring frame and the VLAN information associated with that frame. ISL is currently supported only over Fast Ethernet links, but a single ISL link, or trunk, can carry different protocols from multiple VLANs.

IEEE 802.10 Protocol

The IEEE 802.10 protocol provides connectivity between VLANs. Originally developed to address the growing need for security within shared LAN/MAN environments, it incorporates authentication and encryption techniques to ensure data confidentiality and integrity throughout the network. Additionally, by functioning at Layer 2, it is well suited to high-throughput, low-latency switching environments. IEEE 802.10 protocol can run over any LAN or HDLC serial interface.

ATM LANE Protocol

The ATM LAN Emulation (LANE) protocol provides a way for legacy LAN users to take advantage of ATM benefits without requiring modifications to end-station hardware or software. LANE emulates a broadcast environment like IEEE 802.3 Ethernet on top of an ATM network that is a point-to-point environment.

LAN Emulation makes ATM function like a LAN. LAN Emulation allows standard LAN drivers such as NDIS and ODI to be used. The virtual LAN is transparent to applications. Applications can use normal LAN functions without dealing with the underlying complexities of the ATM implementation. For example, a station can send broadcasts and multicasts, even though ATM is defined as a point-to-point technology and doesn't support any-to-any services.

To accomplish this, special low-level software is implemented on an ATM client workstation, called the *LAN Emulation Client* or *LEC*. The client software communicates with a central control point called a *LAN Emulation Server*, or *LES*. A *Broadcast and Unknown*

Server (*BUS*) acts as a central point to distribute broadcasts and multicasts. The LAN Emulation Configuration Server (LECS) holds a database of LECs. The database is maintained by a network administrator.

VLAN Interoperability

Cisco IOS features bring added benefits to the VLAN technology. Enhancements to ISL, IEEE 802.10, and ATM LAN Emulation (LANE) implementations enable routing of all major protocols between VLANs. These enhancements allow users to create more robust networks, incorporating VLAN configurations by providing communications capabilities between VLANs.

Inter-VLAN Communications

The Cisco IOS supports full routing of several protocols over ISL and ATM LANE virtual LANs. IP, Novell IPX, and AppleTalk routing are supported over IEEE 802.10 VLANs. Standard routing attributes, such as network advertisements, secondaries, and help addresses are applicable and VLAN routing is fast switched. Table 4–1 shows protocols supported for each VLAN encapsulation format and corresponding Cisco IOS releases.

Table 4–1 *Inter-VLAN Routing Protocol Support*

Protocol	ISL	ATM LANE	IEEE 802.10
IP	Release 11.1	Release 10.3	Release 11.1
Novell IPX (default encapsulation)	Release 11.1	Release 10.3	Release 11.1
Novell IPX (configurable encapsulation)	Release 11.3	Release 10.3	Release 11.3
AppleTalk Phase II	Release 11.3	Release 10.3	
DECnet	Release 11.3	Release 11.0	
Banyan VINES	Release 11.3	Release 11.2	
XNS	Release 11.3	Release 11.2	

VLAN Translation

VLAN translation refers to the capability of the Cisco IOS software to translate among different virtual LANs or between VLAN and non-VLAN encapsulating interfaces at Layer 2. Translation is typically used for selective inter-VLAN switching of non-routable protocols and to extend a single VLAN topology across hybrid switching environments. It is also possible to bridge VLANs on the main interface; the VLAN encapsulating header is preserved. Topology changes in one VLAN domain do not affect a different VLAN.

Designing Switched VLANs

By the time you are ready to configure routing between VLANs, you will have already defined them through the switches in your network. Issues related to network design and VLAN definition should be addressed during your network design.

CHAPTER 5

Configuring Routing Between VLANs with ISL Encapsulation

This chapter describes the Inter-Switch Link (ISL) protocol and provides guidelines for configuring ISL features, including the following:

- An Overview of Inter-Switch Link Protocol
- ISL Encapsulation Configuration Task List
- Configuring AppleTalk routing over ISL
- Configuring Banyan VINES routing over ISL
- Configuring DECnet routing over ISL
- Configuring Hot Standby Router Protocol over ISL
- Configuring IPX Routing over ISL
- Configuring VIP Distributed Switching over ISL
- Configuring XNS Routing over ISL
- ISL encapsulation configuration examples

OVERVIEW OF INTER-SWITCH LINK PROTOCOL

Inter-Switch Link (*ISL*) is a Cisco protocol for interconnecting multiple switches and maintaining VLAN information as traffic goes between switches. ISL provides VLAN capabilities while maintaining full-wire–speed performance on Fast Ethernet links in full- or half-duplex mode. ISL operates in a point-to-point environment and will support up to 1000 VLANs. You can define virtually as many logical networks as are necessary for your environment.

Frame Tagging in ISL

With ISL, an Ethernet frame is encapsulated with a header that transports VLAN IDs between switches and routers. A 26-byte header that contains a 10-bit VLAN ID is prepended to the Ethernet frame.

A VLAN ID is added to the frame only when the frame is destined for a non-local network. Figure 5–1 illustrates VLAN packets traversing the shared backbone. Each VLAN packet carries the VLAN ID within the packet header.

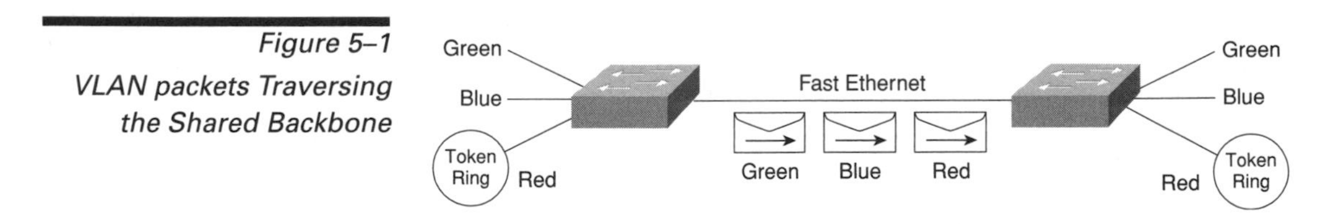

Figure 5–1 VLAN packets Traversing the Shared Backbone

ISL ENCAPSULATION CONFIGURATION TASK LIST

You can configure routing between any number of VLANs in your network. This section documents the configuration tasks for each protocol supported with ISL encapsulation. The basic process is the same, regardless of the protocol being routed. It involves the following:

- Enabling the protocol on the router
- Enabling the protocol on the interface
- Defining the encapsulation format as ISL
- Customizing the protocol according to the requirements for your environment

The configuration processes documented in this chapter include the following:

- Configuring AppleTalk Routing over ISL
- Configuring Banyan VINES Routing over ISL
- Configuring DECnet Routing over ISL
- Configuring Hot Standby Router Protocol over ISL
- Configuring IPX Routing over ISL
- Configuring VIP Distributed Switching over ISL
- Configuring XNS Routing over ISL

Refer to the "ISL Encapsulation Configuration Examples" section at the end of this chapter for sample configurations.

Configuring AppleTalk Routing over ISL

AppleTalk can be routed over virtual LAN (VLAN) subinterfaces using the ISL and IEEE 802.10 VLAN encapsulation protocols. The AppleTalk Routing over ISL and IEEE 802.10 Virtual LANs feature provides full-feature Cisco IOS software AppleTalk support on a per-VLAN basis, allowing standard AppleTalk capabilities to be configured on VLANs.

To route AppleTalk over ISL or IEEE 802.10 between VLANs, you need to customize the subinterface to create the environment in which it will be used. Perform the following tasks in the order in which they appear:

- Enable AppleTalk Routing
- Define the VLAN Encapsulation Format
- Configure AppleTalk on the Subinterface

Enable AppleTalk Routing

To enable AppleTalk routing on either ISL or 802.10 interfaces, perform this task in global configuration mode:

Task	Command
Enable AppleTalk routing globally.	**appletalk routing** [**eigrp** *router-number*]

Define the VLAN Encapsulation Format

To define the VLAN encapsulation format as either ISL or 802.10, perform the following tasks in interface configuration mode:

Task		Command
Step 1	Specify the subinterface the VLAN will use.	**interface** *type slot/port.subinterface-number*
Step 2	Define the encapsulation format as either ISL (**isl**) or IEEE 802.10 (**sde**), and specify the VLAN identifier or security association identifier, respectively.	**encapsulation isl** *vlan-identifier* **encapsulation sde** *said*

Configure AppleTalk on the Subinterface

After you enable AppleTalk globally and define the encapsulation format, you need to enable it on the subinterface by specifying the cable range and naming the AppleTalk zone for each interface. To enable the AppleTalk protocol on the subinterface, perform the following tasks in interface configuration mode:

Task	Command
Assign the AppleTalk cable range and zone for the subinterface.	**appletalk cable-range** *cable-range* [*network.node*]
Assign the AppleTalk zone for the subinterface.	**appletalk zone** *zone-name*

Configuring Banyan VINES Routing over ISL

Banyan VINES can be routed over virtual LAN (VLAN) subinterfaces using the ISL encapsulation protocol. The Banyan VINES Routing over ISL Virtual LANs feature provides full-feature Cisco IOS software Banyan VINES support on a per-VLAN basis, allowing standard Banyan VINES capabilities to be configured on VLANs.

To route Banyan VINES over ISL between VLANs, you need to configure ISL encapsulation on the subinterface. Perform the following tasks in the order in which they appear:

- Enable Banyan VINES Routing
- Define the VLAN Encapsulation Format
- Configure Banyan VINES on the Subinterface

Enable Banyan VINES Routing

To begin the VINES routing configuration, perform this task in global configuration mode:

Task	Command
Enable Banyan VINES routing globally.	**vines routing** [*address*]

Define the VLAN Encapsulation Format

To define the VINES routing encapsulation format, perform these tasks in interface configuration mode:

Task		Command
Step 1	Specify the subinterface on which ISL will be used.	**interface** *type slot/port.subinterface-number*
Step 2	Define the encapsulation format as ISL (**isl**), and specify the VLAN identifier.	**encapsulation isl** *vlan-identifier*

Configure Banyan VINES on the Subinterface

After you enable Banyan VINES globally and define the encapsulation format, you need to enable VINES on the subinterface by specifying the VINES routing metric. To enable the Banyan VINES protocol on the subinterface, perform this task in interface configuration mode:

Task	Command
Enable VINES routing on an interface.	**vines metric** [*whole* [*fractional*]]

CONFIGURING DECNET ROUTING OVER ISL

DECnet can be routed over virtual LAN (VLAN) subinterfaces using the ISL VLAN encapsulation protocols. The DECnet Routing over ISL Virtual LANs feature provides full-feature Cisco IOS software DECnet support on a per-VLAN basis, allowing standard DECnet capabilities to be configured on VLANs.

To route DECnet over ISL VLAN, you need to configure ISL encapsulation on the subinterface. Perform the following tasks in the order in which they appear.

- Enable DECnet Routing
- Define the VLAN Encapsulation Format
- Configure DECnet on the Subinterface

Enable DECnet Routing

To begin the DECnet routing configuration, perform this task in global configuration mode:

Task	Command
Enable DECnet on the router.	**decnet** [*network-number*] **routing** [*decnet-address*]

Define the VLAN Encapsulation Format

To define the encapsulation format, perform these tasks in interface configuration mode:

Task		Command
Step 1	Specify the subinterface on which ISL will be used.	**interface** *type slot/port.subinterface-number*
Step 2	Define the encapsulation format as ISL (**isl**), and specify the VLAN identifier.	**encapsulation isl** *vlan_identifier*

Configure DECnet on the Subinterface

To configure DECnet routing on the subinterface, perform this task in interface configuration mode:

Task	Command
Enable DECnet routing on an interface.	**decnet cost** [*cost-value*]

Configuring Hot Standby Router Protocol over ISL

The *Hot Standby Router Protocol* (*HSRP*) provides fault tolerance and enhanced routing performance for IP networks. HSRP enables Cisco IOS routers to monitor each other's operational status and very quickly assume packet forwarding responsibility in the event the current forwarding device in the HSRP group fails or is taken down for maintenance. The standby mechanism remains transparent to the attached hosts and can be deployed on any LAN type. With multiple hot-standby groups, routers can simultaneously provide redundant backup and perform load-sharing across different IP subnets. Figure 5–2 illustrates HSRP in use with ISL providing routing between several VLANs.

A separate HSRP group is configured for each VLAN subnet so that Cisco IOS Router A can be the primary and forwarding router for VLANs 10 and 20. At the same time, it acts as backup for VLANs 30 and 40. Conversely, Router B acts as the primary and forwarding router for ISL VLANs 30 and 40, as well as the secondary and backup router for distributed VLAN subnets 10 and 20.

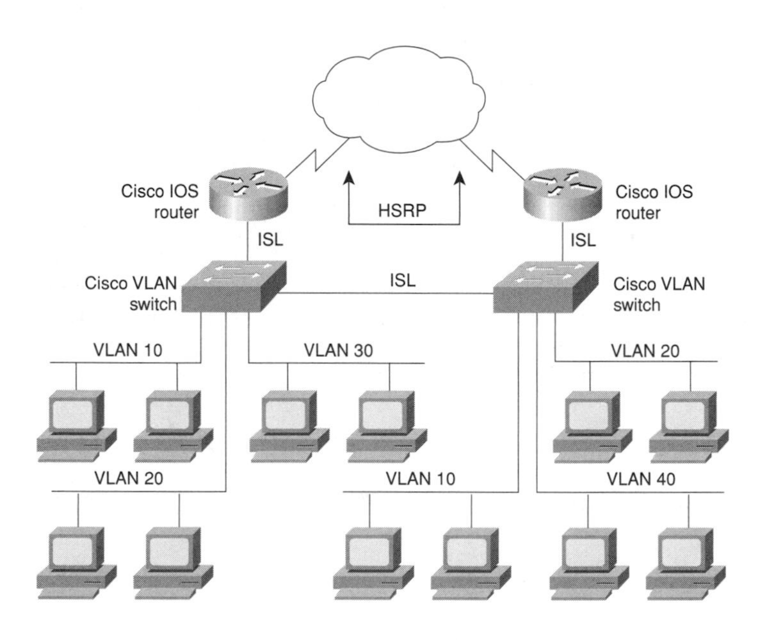

Figure 5–2 Hot Standby Router Protocol in VLAN Configurations

Running HSRP over ISL allows users to configure redundancy between multiple routers that are configured as front ends for VLAN IP subnets. By configuring HSRP over ISLs, users can eliminate situations in which a single point of failure causes traffic interruptions. This feature inherently provides some improvement in overall networking resilience by providing load balancing and redundancy capabilities between subnets and VLANs.

To configure HSRP over ISLs between VLANs, you need to create the environment in which it will be used. Perform the following tasks in the order in which they appear:

- Define the Encapsulation Format
- Define the IP Address
- Enable HSRP

Define the Encapsulation Format

To define the encapsulation format as ISL, perform these tasks in interface configuration mode:

Task		Command
Step 1	Specify the subinterface on which ISL will be used.	**interface** *type slot/port.subinterface-number*
Step 2	Define the encapsulation format, and specify the VLAN identifier.	**encapsulation isl** *vlan-identifier*

Define the IP Address

After you have specified the encapsulation format, define the IP address over which HSRP will be routed. Perform the following task in interface configuration mode:

Task	Command
Specify the IP address for the subnet on which ISL will be used.	**ip address** *ip-address mask* [**secondary**]

Enable HSRP

To enable HSRP on an interface, enable the protocol, and then customize it for the interface. Perform the following task in interface configuration mode:

Task	Command
Enable HSRP.	**standby** [*group-number*] **ip** [*ip-address* [**secondary**]]

To customize "hot standby" group attributes, perform one or more of these tasks in interface configuration mode:

Task	Command
Configure the time between hello packets and the hold time before other routers declare the active router to be down.	**standby** [*group-number*] **timers** *hellotime holdtime*
Set the hot standby priority used to choose the active router.	**standby** [*group-number*] **priority** *priority*

Task	Command
Specify that if the local router has priority over the current active router, the local router should attempt to take its place as the active router.	**standby** [*group-number*] **preempt**
Configure the interface to track other interfaces, so that if one of the other interfaces goes down, the hot standby priority for the device is lowered.	**standby** [*group-number*] **track** *type-number* [*interface-priority*]
Select an authentication string to be carried in all HSRP messages.	**standby** [*group-number*] **authentication** *string*

Configuring IPX Routing over ISL

The IPX Routing over ISL Virtual LANs (VLANs) feature extends Novell NetWare routing capabilities to include support for routing all standard IPX encapsulations for Ethernet frame types in VLAN configurations. Users with Novell NetWare environments can now configure any one of the four IPX Ethernet encapsulations to be routed using the Inter-Switch Link (ISL) encapsulation across VLAN boundaries. IPX encapsulation options now supported for VLAN traffic include the following:

- novell-ether (Novell Ethernet_802.3)
- sap (Novell Ethernet_802.2)
- arpa (Novell Ethernet_II)
- snap (Novell Ethernet_Snap)

NetWare users can now configure consolidated VLAN routing over a single VLAN trunking interface. With configurable Ethernet encapsulation protocols, users have the flexibility of using VLANs regardless of their NetWare Ethernet encapsulation. Configuring Novell IPX encapsulations on a per-VLAN basis facilitates migration between versions of NetWare. NetWare traffic can now be routed across VLAN boundaries with standard encapsulation options (*arpa*, *sap*, and *snap*) previously unavailable.

NOTES

Only one type of IPX encapsulation can be configured per VLAN (subinterface). The IPX encapsulation used must be the same within any particular subnet: a single encapsulation must be used by all NetWare systems that belong to the same virtual LAN.

To configure Cisco IOS software on a router with connected VLANs to exchange different IPX framing protocols, perform the following tasks in the order in which they appear:

- Enable NetWare Routing
- Define the VLAN Encapsulation Format
- Configure NetWare on the Subinterface

Enable NetWare Routing

To enable IPX routing on ISL interfaces, perform the following task in global configuration mode:

Task	Command
Enable IPX routing globally.	**ipx routing** [*node*]

Define the VLAN Encapsulation Format

To define the encapsulation format as ISL, perform the following tasks in interface configuration mode:

Task		Command
Step 1	Specify the subinterface on which ISL will be used.	**interface** *type slot/port.subinterface-number*
Step 2	Define the encapsulation format and specify the VLAN identifier.	**encapsulation isl** *vlan_identifier*

Configure NetWare on the Subinterface

After you enable NetWare globally and define the VLAN encapsulation format, you need to enable the subinterface by specifying the NetWare network number (if necessary) and the encapsulation type. Perform the following task in interface configuration mode:

Task	Command
Specify the IPX encapsulation.	**ipx network** *network* **encapsulation** *encapsulation-type*

NOTES

The default IPX encapsulation format for Cisco IOS routers is "novell-ether" (Novell Ethernet_802.3). If you are running Novell Netware 3.12 or 4.0, the new Novell default encapsulation format is Novell Ethernet_802.2. You should configure the Cisco router with the IPX encapsulation format "sap."

CONFIGURING VIP DISTRIBUTED SWITCHING OVER ISL

With the introduction of the VIP Distributed ISL feature, Inter-Switch Link (ISL) encapsulated IP packets can be switched on Versatile Interface Processor (VIP) controllers installed on Cisco 7500 series routers.

The second generation Versatile Interface Processor (VIP2) provides distributed switching of IP encapsulated in ISL in VLAN configurations. Where an aggregation route performs inter-VLAN routing for multiple VLANs, traffic can be switched autonomously on-card or between cards rather than through the central Route Switch Processor (RSP). Figure 5–3 shows the VIP distributed architecture of the Cisco 7500 series router.

This distributed architecture allows incremental capacity increases by installing additional VIP cards. Using VIP cards for switching the majority of IP VLAN traffic in multiprotocol environments significantly increases routing performance for the other protocols, since the RSP off-loads IP and then can be dedicated to switching the non-IP protocols.

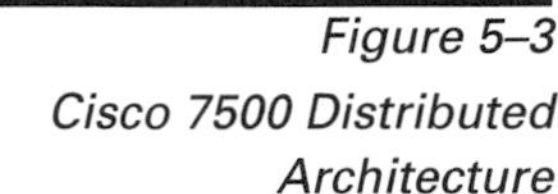

Figure 5–3
Cisco 7500 Distributed Architecture

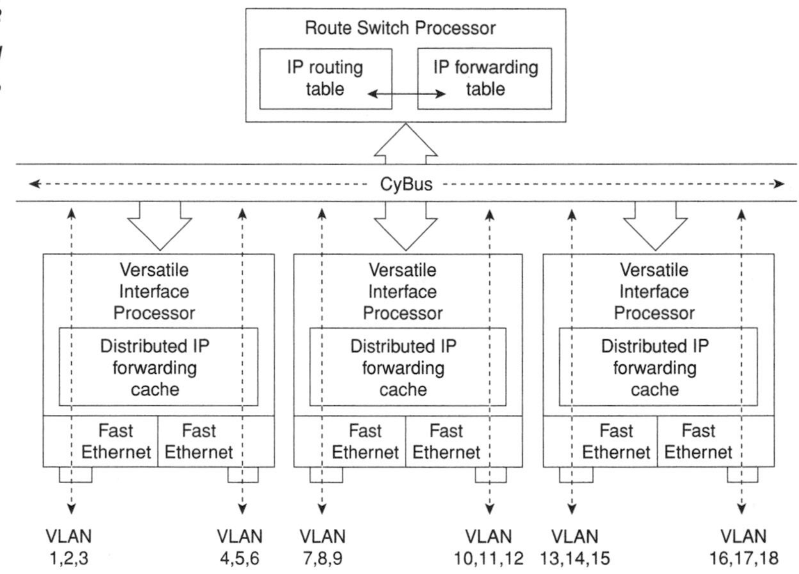

VIP distributed switching offloads switching of ISL VLAN IP traffic to the VIP card, removing involvement from the main CPU. Offloading ISL traffic to the VIP card significantly improves networking performance. Because you can install multiple VIP cards in a router, VLAN routing capacity is increased linearly according to the number of VIP cards installed in the router.

To configure distributed switching on the VIP, you must first configure the router for IP routing. Perform the following tasks in the order in which they appear:

- Enable IP Routing
- Enable VIP Distributed Switching
- Configure ISL Encapsulation on the Subinterface

Enable IP Routing

To enable IP routing, perform the following task in global configuration mode:

Task	Command
Enable IP routing on the router.	**ip routing**

After you have IP routing enabled on the router, you can customize the characteristics to suit your environment.

Enable VIP Distributed Switching

To enable VIP distributed switching, perform the following tasks beginning in interface configuration mode:

Task		Command
Step 1	Specify the interface, and enter interface configuration mode.	**interface** *type slot/port-adapter/port*
Step 2	Enable VIP distributed switching of IP packets on the interface	**ip route-cache distributed**

Configure ISL Encapsulation on the Subinterface

To configure ISL encapsulation on the subinterface, perform the following tasks in interface configuration mode:

Task		Command
Step 1	Specify the interface, and enter interface configuration mode.	**interface** *type slot/port-adapter/port*
Step 2	Define the encapsulation format as ISL and specify the VLAN identifier.	**encapsulation isl** *vlan-identifier*

Configuring XNS Routing over ISL

XNS can be routed over virtual LAN (VLAN) subinterfaces using the ISL VLAN encapsulation protocol. The XNS Routing over ISL Virtual LANs feature provides full-feature Cisco IOS software XNS support on a per-VLAN basis, allowing standard XNS capabilities to be configured on VLANs.

To route XNS over ISL VLANs, you need to configure ISL encapsulation on the subinterface. Perform the following tasks in the order in which they appear:

- Enable XNS Routing
- Define the VLAN Encapsulation Format
- Configure XNS on the Subinterface

Enable XNS Routing

Begin the XNS routing configuration in global configuration mode:

Task	Command
Enable XNS routing globally.	**xns routing** [*address*]

Define the VLAN Encapsulation Format

To define the VLAN encapsulation format, perform the following tasks in interface configuration mode:

Task		Command
Step 1	Specify the subinterface on which ISL will be used.	**interface** *type slot /port.subinterface-number*
Step 2	Define the encapsulation format as ISL (**isl**), and specify the VLAN identifier.	**encapsulation isl** *vlan-identifier*

Configure XNS on the Subinterface

Enable XNS on the subinterface by specifying the XNS network number. Perform the following task in interface configuration mode:

Task	Command
Enable XNS routing on the subinterface	**xns network** [*number*]

ISL Encapsulation Configuration Examples

This section provides configuration examples for each of the protocols described in this chapter. It includes the following examples:

- AppleTalk Routing over ISL Configuration Examples
- Banyan VINES Routing over ISL Configuration Example
- DECnet Routing over ISL Configuration Example
- HSRP over ISL Configuration Example
- IPX Routing over ISL Configuration Example
- VIP Distributed Switching over ISL Configuration Example
- XNS Routing over ISL Configuration Example

AppleTalk Routing over ISL Configuration Examples

The configuration example illustrated in Figure 5–4 shows AppleTalk being routed between different ISL and IEEE 802.10 VLAN encapsulating subinterfaces.

As shown in Figure 5–4, AppleTalk traffic is routed to and from switched VLAN domains 3, 4, 100, and 200 to any other AppleTalk routing interface. This example shows a sample configuration file for the Cisco 7500 series router with the commands entered to configure the network shown in Figure 5–4.

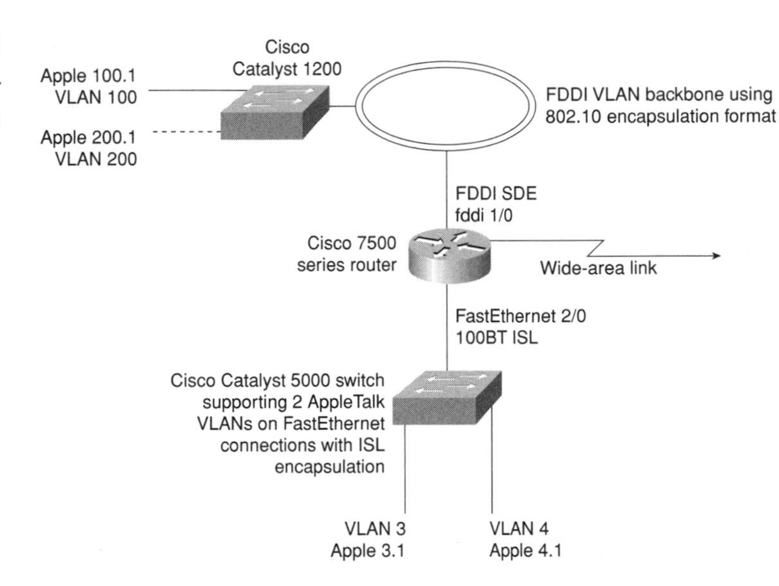

Figure 5–4
Routing AppleTalk over VLAN Encapsulations

Configuration for the Cisco 7500 Router

```
!
appletalk routing
interface Fddi 1/0.100
 encapsulation sde 100
 appletalk cable-range 100-100 100.2
 appletalk zone 100
!
interface Fddi 1/0.200
 encapsulation sde 200
 appletalk cable-range 200-200 200.2
 appletalk zone 200
!
interface FastEthernet 2/0.3
 encapsulation isl 3
 appletalk cable-range 3-3 3.2
 appletalk zone 3
!
interface FastEthernet 2/0.4
 encapsulation isl 4
 appletalk cable-range 4-4 4.2
 appletalk zone 4
!
```

Banyan VINES Routing over ISL Configuration Example

To configure routing of the Banyan VINES protocol over ISL trunks, you need to define ISL as the encapsulation type. This example shows Banyan VINES configured to be routed over an ISL trunk:

Example Banyan VINES Configuration

```
vines routing
interface fastethernet 0.1
 encapsulation isl 100
 vines metric 2
```

DECnet Routing over ISL Configuration Example

To configure routing the DECnet protocol over ISL trunks, you need to define ISL as the encapsulation type. The following example shows DECnet configured to be routed over an ISL trunk:

Example DECnet Configuration

```
decnet routing 2.1
interface fastethernet 1/0.1
 encapsulation isl 200
 decnet cost 4
```

HSRP over ISL Configuration Example

The configuration example shown in Figure 5–5 shows HSRP being used on two VLAN routers sending traffic to and from ISL VLANs through a Catalyst 5000 switch. Each router forwards its own traffic and acts as a standby for the other.

The topology shown in Figure 5–5 illustrates a Cisco Catalyst VLAN switch supporting Fast Ethernet connections to two routers running HSRP. Both routers are configured to route HSRP over ISLs.

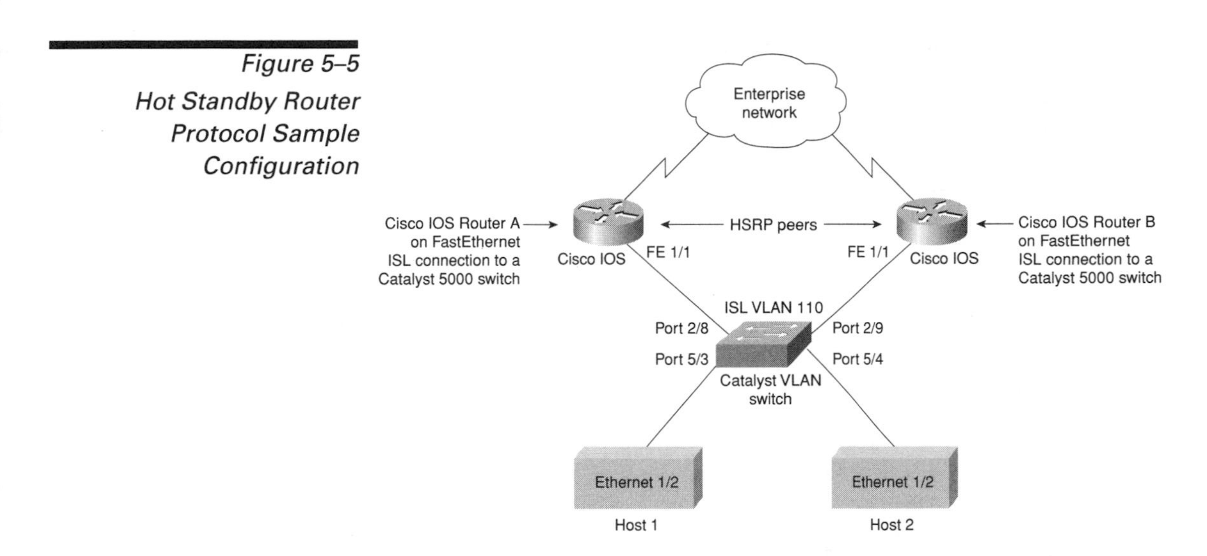

Figure 5–5
Hot Standby Router Protocol Sample Configuration

The standby conditions are determined by the standby commands used in the configuration. Traffic from Host 1 is forwarded through Router A. Because the priority for the group is higher, Router A is the active router for Host 1. Because the priority for the group serviced by Host 2 is higher in Router B, traffic from Host 2 is forwarded through Router B, making Router B its active router.

In the configuration shown in Figure 5–5, if the active router becomes unavailable, the standby router assumes active status for the additional traffic and automatically routes the traffic normally handled by the router that has become unavailable.

Host 1 Configuration

```
interface Ethernet 1/2
 ip address 110.1.1.25 255.255.255.0
 ip route 0.0.0.0 0.0.0.0 110.1.1.101
```

Host 2 Configuration

```
interface Ethernet 1/2
 ip address 110.1.1.27 255.255.255.0
 ip route 0.0.0.0 0.0.0.0 110.1.1.102
!
```

Router A Configuration

```
interface FastEthernet 1/1.110
 encapsulation isl 110
```

```
 ip address 110.1.1.2 255.255.255.0
 standby 1 ip 110.1.1.101
 standby 1 preempt
 standby 1 priority 105
 standby 2 ip 110.1.1.102
 standby 2 preempt

!
end

!
```

Router B Configuration

```
interface FastEthernet 1/1.110
 encapsulation isl 110
 ip address 110.1.1.3 255.255.255.0
 standby 1 ip 110.1.1.101
 standby 1 preempt
 standby 2 ip 110.1.1.102
 standby 2 preempt
 standby 2 priority 105
router igrp 1
!
network 110.1.0.0
network 120.1.0.0
!
```

VLAN Switch Configuration

```
set vlan 110 5/4
set vlan 110 5/3
set trunk 2/8 110
set trunk 2/9 110
```

IPX Routing over ISL Configuration Example

Figure 5–6 shows IPX interior encapsulations configured over ISL encapsulation in VLAN configurations. Note that three different IPX encapsulation formats are used. VLAN 20 uses sap encapsulation, VLAN 30 uses arpa, and VLAN 70 uses novell-ether encapsulation. Prior to the introduction of this feature, only the default encapsulation format, "novell-ether," was available for routing IPX over ISLlinks in VLANs.

Figure 5–6
Configurable IPX Encapsulations Routed over ISL in VLAN Configurations

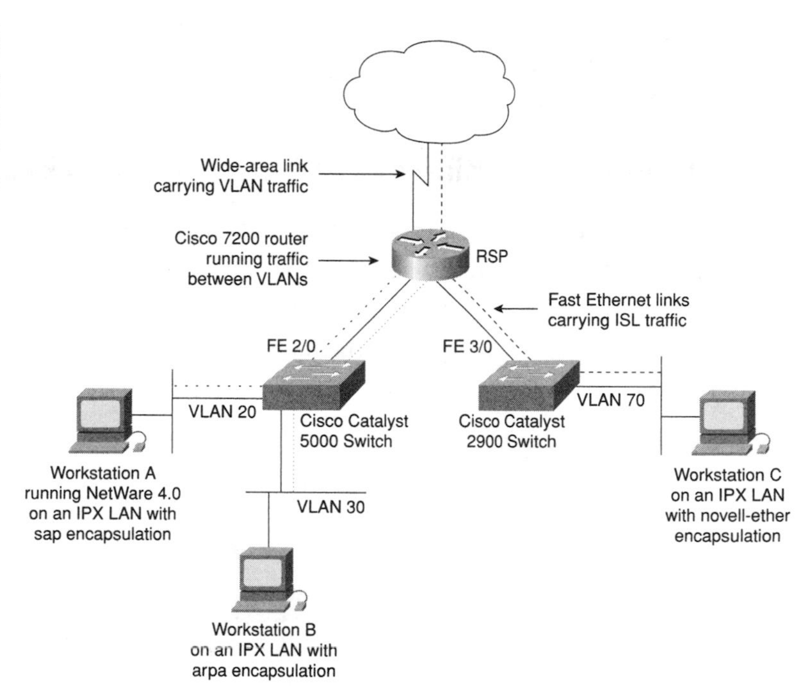

VLAN 20 Configuration

```
ipx routing
interface FastEthernet 2/0
 no shutdown
interface FastEthernet 2/0.20
 encapsulation isl 20
 ipx network 20 encapsulation sap
```

VLAN 30 Configuration

```
ipx routing
interface FastEthernet 2/0
 no shutdown
interface FastEthernet 2/0.30
 encapsulation isl 30
 ipx network 30 encapsulation arpa
```

VLAN 70

```
ipx routing
interface FastEthernet 3/0
 no shutdown
```

```
interface Fast3/0.70
 encapsulation isl 70
 ipx network 70 encapsulation novell-ether
```

VIP Distributed Switching over ISL Configuration Example

Figure 5–7 illustrates a topology in which Catalyst VLAN switches are connected to routers forwarding traffic from a number of ISL VLANs. With the VIP distributed ISL capability in the Cisco 7500 series router, each VIP card can route ISL-encapsulated VLAN IP traffic. The inter-VLAN routing capacity is increased linearly by the packet-forwarding capability of each VIP card.

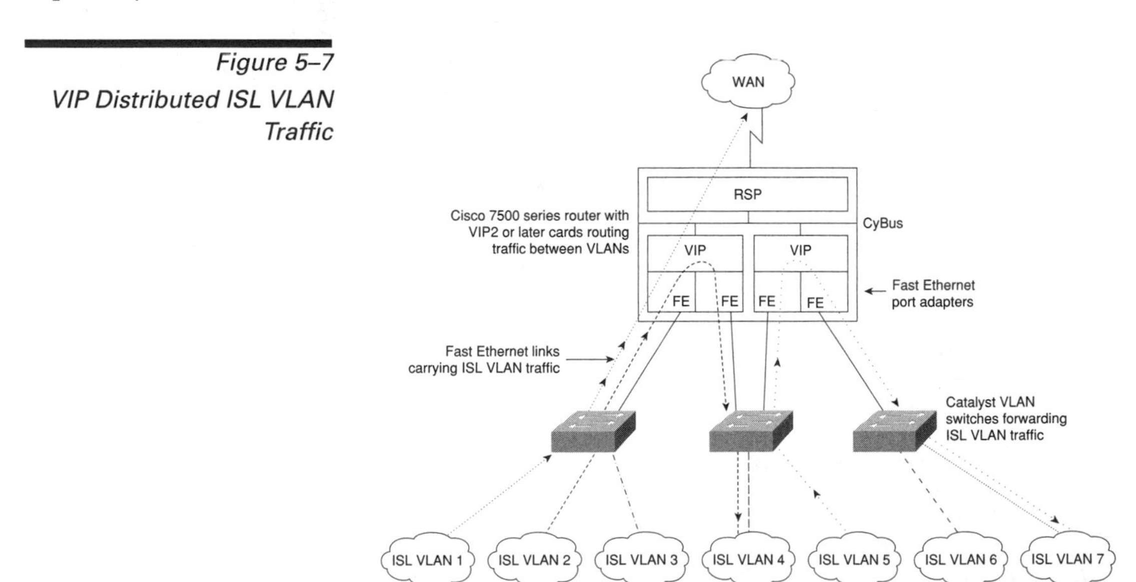

Figure 5–7
VIP Distributed ISL VLAN Traffic

In Figure 5–7, the VIP cards forward the traffic between ISL VLANs or any other routing interface. Traffic from any VLAN can be routed to any of the other VLANs, regardless of which VIP card receives the traffic.

The following commands show the configuration for each of the VLANs shown in Figure 5–7:

```
interface FastEthernet1/0/0
 ip address 20.1.1.1 255.255.255.0
 ip route-cache distributed
 full-duplex
```

```
interface FastEthernet1/0/0.1
 ip address 22.1.1.1 255.255.255.0
 encapsulation isl 1

interface FastEthernet1/0/0.2
 ip address 22.1.2.1 255.255.255.0
 encapsulation isl 2

interface FastEthernet1/0/0.3
 ip address 22.1.3.1 255.255.255.0
 encapsulation isl 3

interface FastEthernet1/1/0
 ip route-cache distributed
 full-duplex

interface FastEthernet1/1/0.1
 ip address 77.1.1.1 255.255.255.0
 encapsulation isl 4

interface Fast Ethernet 2/0/0
 ip address 30.1.1.1 255.255.255.0
 ip route-cache distributed
 full-duplex

interface FastEthernet2/0/0.5
 ip address 33.1.1.1 255.255.255.0
 encapsulation isl 5

interface FastEthernet2/1/0
 ip address 40.1.1.1 255.255.255.0
 ip route-cache distributed
 full-duplex

interface FastEthernet2/1/0.6
 ip address 44.1.6.1 255.255.255.0
 encapsulation isl 6

interface FastEthernet2/1/0.7
 ip address 44.1.7.1 255.255.255.0
 encapsulation isl 7
```

XNS Routing over ISL Configuration Example

To configure routing of the XNS protocol over ISL trunks, you need to define ISL as the encapsulation type. The following example shows XNS configured to be routed over an ISL trunk:

Example XNS Configuration

```
xns routing 0123.4567.adcb
interface fastethernet 1/0.1
 encapsulation isl 100
 xns network 20
```

CHAPTER 6

Configuring Routing Between VLANs with IEEE 802.10 Encapsulation

The IEEE 802.10 standard provides a method for secure bridging of data across a shared backbone. It defines a single frame type known as the Secure Data Exchange (SDE), a MAC-layer frame with an IEEE 802.10 header inserted between the MAC header and the frame data. A well-known Logical Link Control Service Access Point notifies the switch of an incoming IEEE 802.10 frame. The VLAN ID is carried in the 4-byte Security Association Identifier (SAID) field.

NOTES

HDLC Serial links can be used as VLAN trunks in IEEE 802.10 virtual LANs to extend a virtual topology beyond a LAN backbone.

CONFIGURING APPLETALK ROUTING OVER IEEE 802.10

AppleTalk can be routed over virtual LAN (VLAN) subinterfaces using the ISL or IEEE 802.10 VLAN encapsulation protocols. The AppleTalk Routing over IEEE 802.10 Virtual LANs feature provides full-feature Cisco IOS software AppleTalk support on a per-VLAN basis, allowing standard AppleTalk capabilities to be configured on VLANs.

AppleTalk users can now configure consolidated VLAN routing over a single VLAN trunking interface. Prior to introduction of this feature, AppleTalk could be routed only on the main interface on a LAN port. If AppleTalk routing was disabled on the main interface or if the main interface was shut down, the entire physical interface would stop routing any AppleTalk packets. With this feature enabled, AppleTalk routing on subinterfaces will be unaffected by changes in the main interface's "no-shut" state.

To route AppleTalk over IEEE 802.10 between VLANs, you need to create the environment in which it will be used by customizing the subinterface. Perform the following tasks in the order in which they appear:

- Enable AppleTalk Routing
- Configure AppleTalk on the Subinterface

Enable AppleTalk Routing

To enable AppleTalk routing on IEEE 802.10 interfaces, perform the following task in global configuration mode:

Task	Command
Enable AppleTalk routing globally.	**appletalk routing** [**eigrp** *router-number*]

Configure AppleTalk on the Subinterface

After you enable AppleTalk globally and define the encapsulation format, you need to enable it on the subinterface by specifying the cable range and naming the AppleTalk zone for each interface. To enable the AppleTalk protocol on the subinterface, perform the following tasks in interface configuration mode:

Task	Command
Assign the AppleTalk cable range and zone for the subinterface.	**appletalk cable-range** *cable-range* [*network.node*]
Assign the AppleTalk zone for the subinterface.	**appletalk zone** *zone-name*

Configuration Example for Routing AppleTalk over IEEE 802.10

The configuration example illustrated in Figure 6–1 shows AppleTalk being routed between different ISL and IEEE 802.10 VLAN encapsulating subinterfaces.

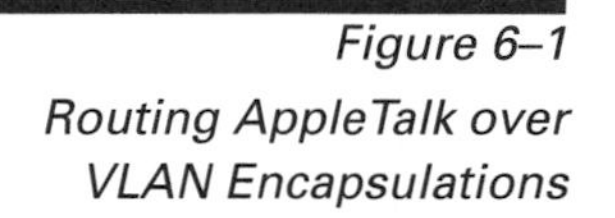

Figure 6–1
Routing AppleTalk over VLAN Encapsulations

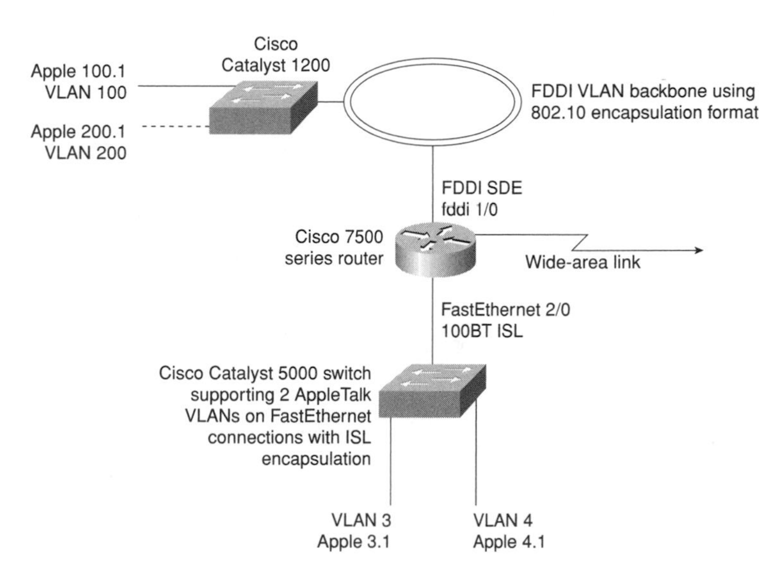

As shown in Figure 6–1, AppleTalk traffic is routed to and from switched VLAN domains 3, 4, 100, and 200 to any other AppleTalk routing interface. This example shows a sample configuration file for the Cisco 7500 series router with the commands entered to configure the network shown in Figure 6–1.

Configuration for the Cisco 7500 Router

```
!
interface Fddi 1/0.100
 encapsulation sde 100
 appletalk cable-range 100-100 100.2
 appletalk zone 100
!
interface Fddi 1/0.200
 encapsulation sde 200
 appletalk cable-range 200-200 200.2
 appletalk zone 200
!
interface FastEthernet 2/0.3
 encapsulation isl 3
 appletalk cable-range 3-3 3.2
 appletalk zone 3
!
interface FastEthernet 2/0.4
 encapsulation isl 4
 appletalk cable-range 4-4 4.2
 appletalk zone 4
!
```

CHAPTER 7

LAN Emulation Overview

This overview chapter gives a high-level description of LAN Emulation (LANE). For specific configuration information, refer to Chapter 8, "Configuring LAN Emulation."

LAN EMULATION (LANE)

Cisco's implementation of LANE makes an ATM interface look like one or more Ethernet interfaces. LANE is an ATM service defined by the ATM Forum specification *LAN Emulation over ATM*, ATM_FORUM 94-0035. This service emulates the following LAN-specific characteristics:

- Connectionless services
- Multicast services
- LAN media access control (MAC) driver services

LANE service provides connectivity between ATM-attached devices and connectivity with LAN-attached devices. This includes connectivity between ATM-attached stations and LAN-attached stations, and also connectivity between LAN-attached stations across an ATM network.

Because LANE connectivity is defined at the MAC layer, upper protocol layer functions of LAN applications can continue unchanged when the devices join emulated LANs. This feature protects corporate investments in legacy LAN applications.

An ATM network can support multiple independent emulated LAN networks. Membership of an end system in any of the emulated LANs is independent of the physical location of the end system. This characteristic enables easy hardware moves and location changes. In addition, the end systems can also move easily from one emulated LAN to another, whether or not the hardware moves.

LAN emulation in an ATM environment provides routing between emulated LANs for supported routing protocols and high-speed, scalable switching of local traffic.

The ATM LANE system has three servers that are single points of failure. These are the LECS (Configuration Server), the LES (emulated LAN server), and the BUS (the broadcast and unknown server). Beginning with Release 11.2, LANE fault tolerance or Simple LANE Service Replication on the emulated LAN provides backup servers to prevent problems if these servers fail.

The fault tolerance mechanism that eliminates these single points of failure is described in Chapter 8. Although this scheme is proprietary, no new protocol additions have been made to the LANE subsystems.

LANE Components

Any number of emulated LANs can be set up in an ATM switch cloud. A router can participate in any number of these emulated LANs.

LANE is defined on a LAN client-server model. The following components are implemented in this release:

- **LANE client:** A LANE client emulates a LAN interface to higher layer protocols and applications. It forwards data to other LANE components and performs LANE address resolution functions. Each LANE client is a member of only one emulated LAN. However, a router can include LANE clients for multiple emulated LANs: one LANE client for *each* emulated LAN of which it is a member. If a router has clients for multiple emulated LANs, the Cisco IOS software can route traffic between the emulated LANs.

- **LANE server:** The LANE server for an emulated LAN is the control center. It provides joining, address resolution, and address registration services to the LANE clients in that emulated LAN. Clients can register destination unicast and multicast MAC addresses with the LANE server. The LANE server also handles LANE ARP (LE ARP) requests and responses. Our implementation has a limit of one LANE server per emulated LAN.
- **LANE broadcast-and-unknown server:** The LANE broadcast-and-unknown server sequences and distributes multicast and broadcast packets and handles unicast flooding. In this release, the LANE server and the LANE broadcast-and-unknown server are combined and located in the same Cisco 7000 family or Cisco 4500 series router: one combined LANE server and broadcast-and-unknown server is required per emulated LAN.
- **LANE configuration server:** The LANE configuration server contains the database that determines to which emulated LAN a device belongs (each configuration server can have a different named database). Each LANE client consults the LANE configuration server just once, when it joins an emulated LAN, to determine which emulated LAN it should join. The LANE configuration server returns the ATM address of the LANE server for that emulated LAN. One LANE configuration server is required per LANE ATM switch cloud.

  The LANE configuration server's database can have the following four types of entries:

  - Emulated LAN name-ATM address of LANE server pairs
  - LANE client MAC address-emulated LAN name pairs
  - LANE client ATM template-emulated LAN name pairs
  - Default emulated LAN name

NOTES

Emulated LAN names must be unique on an interface. If two interfaces participate in LANE, the second interface may be in a different switch cloud.

LANE Operation and Communication

Communication among LANE components is ordinarily handled by several types of switched virtual circuits (SVCs). Some SVCs are unidirectional while others are bidirectional. Some are point-to-point and others are point-to-multipoint. Figure 7–1 illustrates the various virtual channel connections (VCCs)—also known as *virtual circuit connections*—that are used in LANE configuration. In this figure, *LE server* stands for the LANE server, *LECS* stands for the LANE configuration server, and *BUS* stands for the LANE broadcast-and-unknown server.

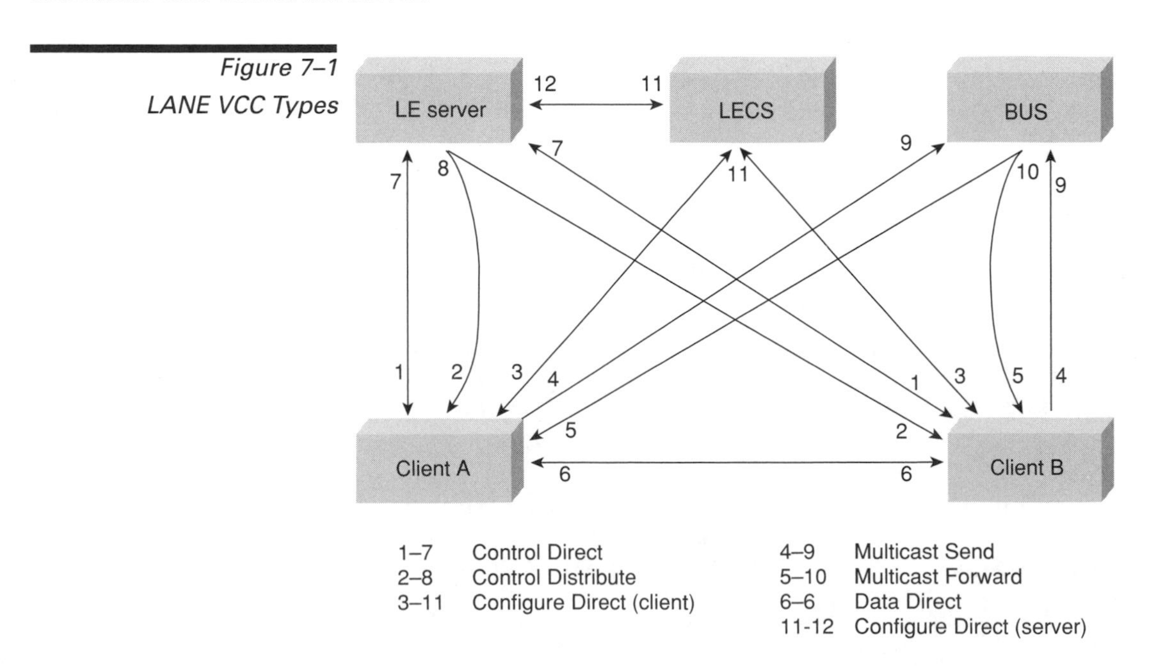

Figure 7–1
LANE VCC Types

The following section describes various processes that occur, starting with a client requesting to join an emulated LAN after the component routers have been configured.

Client Joining a Emulated LAN

The following process normally occurs after a LANE client has been enabled:

- **Client requests to join an emulated LAN:** The client sets up a connection to the LANE configuration server—a bidirectional point-to-point Configure Direct VCC—to find the ATM address of the LANE server for its emulated LAN. LANE clients find the LANE configuration server by using the following methods in the listed order:
    - Locally configured ATM address
    - Interim Local Management Interface (ILMI)
    - Fixed address defined by the ATM Forum
    - PVC 0/17
- **Configuration server identifies the LANE server:** Using the same VCC, the LANE configuration server returns the ATM address and the name of the LANE server for the client's emulated LAN.
- **Client contacts the server for its LAN:** The client sets up a connection to the LANE server for its emulated LAN (a bidirectional point-to-point Control Direct VCC) to exchange control traffic. After a Control Direct VCC is established between a LANE client and a LANE server, it remains up.
- **Server verifies that the client is allowed to join the emulated LAN:** The server for the emulated LAN sets up a connection to the LANE configuration server to verify that the client is allowed to join the emulated LAN—a bidirectional point-to-point Configure Direct (server) VCC. The server's configuration request contains the client's MAC address, its ATM address, and the name of the emulated LAN. The LANE configuration server checks its database to determine whether the client can join that LAN; then it uses the same VCC to inform the server whether the client is or is not allowed to join.
- **LANE server allows or disallows the client to join the emulated LAN:** If allowed, the LANE server adds the LANE client to the unidirectional point-to-multipoint Control Distribute VCC and confirms the join over the bidirectional

point-to-point Control Direct VCC. If disallowed, the LANE server rejects the join over the bidirectional point-to-point Control Direct VCC.

- **LANE client sends LE ARP packets for the broadcast address, which is all 1s:** Sending LE ARP packets for the broadcast address sets up the VCCs to and from the broadcast-and-unknown server.

Address Resolution

As communication occurs on the emulated LAN, each client dynamically builds a local LANE ARP (LE ARP) table. A client's LE ARP table can also have static, preconfigured entries. The LE ARP table maps MAC addresses to ATM addresses.

NOTES

LE ARP is not the same as IP ARP. IP ARP maps IP addresses (Layer 3) to Ethernet MAC addresses (Layer 2); LE ARP maps emulated LAN MAC addresses (Layer 2) to ATM addresses (also Layer 2).

When a client first joins an emulated LAN, its LE ARP table has no dynamic entries and the client has no information about destinations on or behind its emulated LAN. To learn about a destination when a packet is to be sent, the client begins the following process to find the ATM address corresponding to the known MAC address:

1. The client sends an LE ARP request to the LANE server for this emulated LAN (point-to-point Control Direct VCC).
2. The LANE server forwards the LE ARP request to all clients on the emulated LAN (point-to-multipoint Control Distribute VCC).
3. Any client that recognizes the MAC address responds with its ATM address (point-to-point Control Direct VCC).
4. The LANE server forwards the response (point-to-multipoint Control Distribute VCC).
5. The client adds the MAC address-ATM address pair to its LE ARP cache.
6. Then the client can establish a VCC to the desired destination and transmit packets to that ATM address (bidirectional point-to-point Data Direct VCC).

For unknown destinations, the client sends a packet to the broadcast-and-unknown server, which forwards the packet to all clients via flooding. The broadcast-and-unknown server floods the packet because the destination might be behind a bridge that has not yet learned this particular address.

Multicast Traffic

When a LANE client has broadcast or multicast traffic, or unicast traffic with an unknown address to send, the following process occurs:

1. The client sends the packet to the broadcast-and-unknown server (unidirectional point-to-point Multicast Send VCC).
2. The broadcast-and-unknown server forwards (floods) the packet to all clients (unidirectional point-to-multipoint Multicast Forward VCC). This VCC branches at each ATM switch. The switch forwards such packets to multiple outputs. (The switch does not examine the MAC addresses; it simply forwards all packets it receives.)

Typical LANE Scenarios

In typical LANE cases, one or more Cisco 7000 family routers, or Cisco 4500 series routers, are attached to a Cisco LightStream ATM switch. The LightStream ATM switch provides connectivity to the broader ATM network switch cloud. The routers are configured to support one or more emulated LANs. One of the routers is configured to perform the LANE configuration server functions. A router is configured to perform the server function and the broadcast-and-unknown server function for each emulated LAN. (One router can perform the server function and the broadcast-and-unknown server function for several emulated LANs.) In addition to these functions, each router also acts as a LANE client for one or more emulated LANs.

This section presents two scenarios using the same four Cisco routers and the same Cisco LightStream ATM switch. Figure 7–2 illustrates a scenario in which one emulated LAN is set up on the switch and routers. Figure 7–3 illustrates a scenario in which several emulated LANs are set up on the switch and routers.

The physical layout and the physical components of an emulated network might not differ for the single and the multiple emulated LAN cases. The differences are in the software configuration for the number of emulated LANs and the assignment of LANE components to the different physical components.

Single Emulated LAN Scenario

In a single emulated LAN scenario, the LANE components might be assigned as follows:

- Router 1 includes the following LANE components:
    - The LANE configuration server (one per LANE switch cloud)
    - The LANE server and broadcast-and-unknown server for the emulated LAN with the default name *man* (for Manufacturing)
    - The LANE client for the *man* emulated LAN
- Router 2 includes a LANE client for the *man* emulated LAN.
- Router 3 includes a LANE client for the *man* emulated LAN.
- Router 4 includes a LANE client for the *man* emulated LAN.

Figure 7–2 illustrates this single emulated LAN configured across several routers.

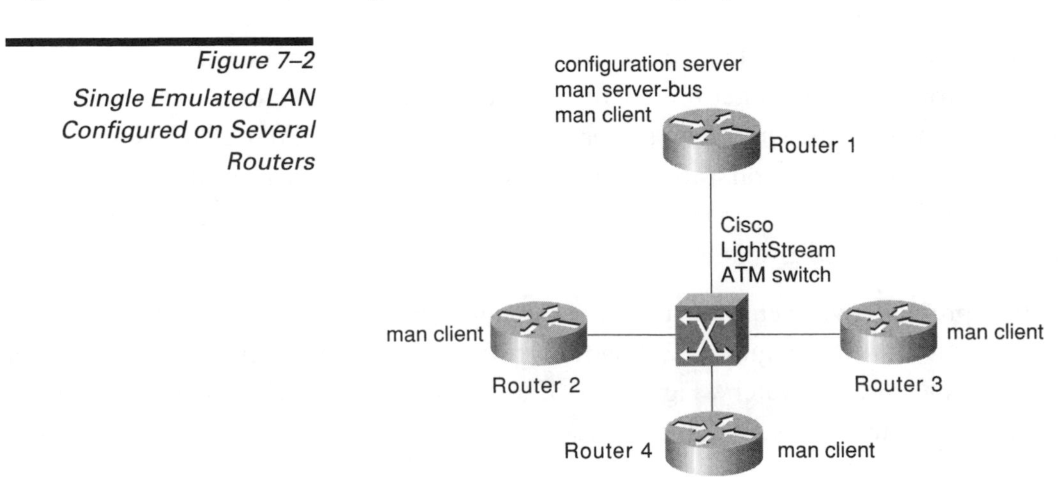

Figure 7–2 Single Emulated LAN Configured on Several Routers

Multiple Emulated LAN Scenario

In the multiple LAN scenario, the same switch and routers are used but multiple emulated LANs are configured (see Figure 7–3).

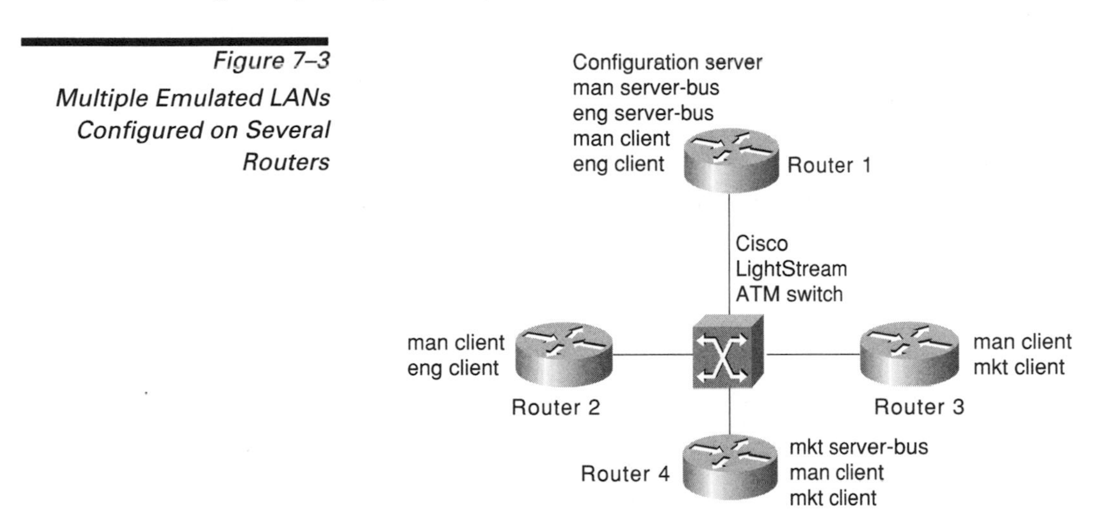

Figure 7–3
Multiple Emulated LANs Configured on Several Routers

In the following scenario, three emulated LANs are configured on four routers:

- Router 1 includes following LANE components:
    - The LANE configuration server (one per LANE switch cloud)
    - The LANE server and broadcast-and-unknown server for the emulated LAN called *man* (for Manufacturing)
    - The LANE server and broadcast-and-unknown server functions for the emulated LAN called *eng* (for Engineering)
    - A LANE client for the *man* emulated LAN
    - A LANE client for the *eng* emulated LAN
- Router 2 includes only the LANE clients for the *man* and *eng* emulated LANs.
- Router 3 includes only the LANE clients for the *man* and *mkt* (for Marketing) emulated LANs.

- Router 4 includes the following LANE components:
  - The LANE server and broadcast-and-unknown server for the *mkt* emulated LAN
  - A LANE client for the *man* emulated LAN
  - A LANE client for the *mkt* emulated LANs

In this scenario, once routing is enabled and network level addresses are assigned, Router 1 and Router 2 can route between the *man* and the *eng* emulated LANs, and Router 3 and Router 4 can route between the *man* and the *mkt* emulated LANs.

CHAPTER 8

Configuring LAN Emulation

This chapter describes how to configure LAN emulation (LANE) on the following platforms that are connected to an Asynchronous Transfer Mode (ATM) switch or switch cloud:

- ATM Interface Processor (AIP) on the Cisco 7500 series routers
- ATM port adapter on the Cisco 7200 series and Cisco 7500 series routers
- Network Processor Module (NPM) on the Cisco 4500 and Cisco 4700 routers

NOTES

In Cisco IOS Release 11.3, all commands supported on the Cisco 7500 series routers are also supported on the Cisco 7000 series.

LANE ON ATM

LANE emulates an IEEE 802.3 Ethernet or IEEE 802.5 Token Ring LAN using ATM technology. LANE provides a service interface for network layer protocols that is identical to existing MAC layers. No changes are required to existing upper layer protocols and applications. With LANE, Ethernet and Token Ring packets are encapsulated in the appropriate

ATM cells and sent across the ATM network. When the packets reach the other side of the ATM network, they are de-encapsulated. LANE essentially bridges LAN traffic across ATM switches.

Benefits of LANE

ATM is a cell-switching and multiplexing technology designed to combine the benefits of circuit switching (constant transmission delay and guaranteed capacity) with those of packet switching (flexibility and efficiency for intermittent traffic).

LANE allows legacy Ethernet and Token Ring LAN users to take advantage of ATM's benefits without modifying end-station hardware or software. ATM uses connection-oriented service with point-to-point signaling or multicast signaling between source and destination devices. However, LANs use connectionless service. Messages are broadcast to all devices on the network. With LANE, routers and switches emulate the connectionless service of a LAN for the endstations.

By using LANE, you can scale your networks to larger sizes while preserving your investment in LAN technology.

LANE Components

A single emulated LAN consists of the following entities: A LANE configuration server, a broadcast-and-unknown server, a LANE server, and LANE clients.

- **LANE configuration server:** A server that assigns individual clients to particular emulated LANs by directing them to the LANE server for the emulated LAN. The LANE configuration server (LECS) maintains a database of LANE client and server ATM or MAC addresses and their emulated LANs. An LECS can serve multiple emulated LANs.
- **broadcast-and-unknown server:** A multicast server that floods unknown destination traffic and forwards multicast and broadcast traffic to clients within an emulated LAN. One broadcast-and-unknown server (BUS) exists per emulated LAN.

- **LANE server:** A server that provides a registration facility for clients to join the emulated LAN. There is one LANE server (LES) per emulated LAN. The LANE server handles LAN Emulation Address Resolution Protocol (LE ARP) requests and maintains a list of LAN destination MAC addresses. For Token Ring LANE, the LANE server also maintains a list of route-descriptors that is used to support source-route bridging over the emulated LAN. The route-descriptors are used to determine the ATM address of the next hop in the Routing Information Field (RIF).
- **LANE client:** An entity in an endpoint, such as a router, that performs data forwarding, address resolution, and other control functions for a single endpoint in a single emulated LAN. The LANE client (LEC) provides standard LAN service to any higher layers that interface with it. A router can have multiple resident LANE clients, each connecting with different emulated LANs. The LANE client registers its MAC and ATM addresses with the LANE server.

Emulated LAN entities coexist on one or more Cisco routers. On Cisco routers, the LANE server and the BUS are combined into a single entity. Other LANE components include ATM switches; that is, any ATM switch that supports the Interim Local Management Interface (ILMI) and signaling. Multiple emulated LANs can coexist on a single ATM network.

Simple Server Redundancy

LANE relies on three servers: the LANE configuration server, the LANE server, and the BUS. If any one of these servers fails, the emulated LAN cannot fully function.

Cisco has developed a fault tolerance mechanism known as *simple server redundancy* that eliminates these single points of failure. Although this scheme is proprietary, no new protocol additions have been made to the LANE subsystems.

Simple server redundancy uses multiple LANE configuration servers and multiple broadcast-and-unknown and LANE servers. You can configure servers as backup servers, which will become active if a master server fails. The priority levels for the servers determine which servers have precedence.

IMPLEMENTATION CONSIDERATIONS

The following sections contain information relevant to implementation:

- Network Support
- Hardware Support
- Addressing
- Rules for Assigning Components to Interfaces and Subinterfaces

Network Support

In this release, Cisco supports the following networking features:

- Ethernet-emulated LANs
    - Routing from one emulated LAN to another via IP, IPX, or AppleTalk
    - Bridging between emulated LANs and between emulated LANs and other LANs
    - DECnet, Banyan VINES, and XNS routed protocols
- Token-Ring emulated LANs
    - IP routing (fast switched) between emulated LANs and between a Token Ring emulated LAN and a legacy LAN
    - IPX routing between emulated LANs and between a Token Ring emulated LAN and a legacy LAN
    - Two-port and multiport source-route bridging (fast switched) between emulated LANs and between emulated LANs and a Token Ring
    - IP and IPX multiring
    - Source-route bridging (SRB), source-route translational bridging (SR/TLB), and source-route transparent bridging (SRT)
    - AppleTalk, DECnet, Banyan VINES, and XNS protocols are not supported

NOTES

Cisco's implementation of LAN Emulation over 802.5 uses existing terminology and configuration options for Token Rings, including source-route bridging. Transparent bridging and Advanced Peer-to-Peer Networking (APPN) are not supported at this time.

- Hot Standby Router Protocol (HSRP)

Hardware Support

This release of LANE is supported on the following platforms:

- Cisco 4500-M, Cisco 4700-M
- Cisco 7200 series
- Cisco 7500 series

NOTES

In Cisco IOS Release 11.3, all commands supported on the Cisco 7500 series routers are also supported on the Cisco 7000 series routers equipped with RSP7000. Token Ring LAN emulation on Cisco 7000 series routers requires the RSP7000 upgrade. The RSP7000 upgrade requires a minimum of 24 MB DRAM and 8 MB Flash memory.

The router must contain an ATM Interface Processor (AIP), ATM port adapter, or an NP-1A ATM Network Processor Module (NPM). These modules provide an ATM network interface for the routers. Network interfaces reside on modular interface processors, which provide a direct connection between the high-speed Cisco Extended Bus (CxBus) and the external networks. The maximum number of AIPs, ATM port adapters, or NPMs that the router supports depends on the bandwidth configured. The total bandwidth through all the AIPs, ATM port adapters, or NPMs in the system should be limited to 200 Mbps full duplex—two Transparent Asynchronous Transmitter/Receiver Interfaces (TAXIs), one Synchronous Optical Network (SONET) and one E3, or one SONET and one lightly used SONET.

This feature also requires one of the following switches:

- Cisco LightStream 1010 (recommended)
- Cisco LightStream 100
- Any ATM switch with UNI 3.0/3.1 and ILMI support for communicating the LECS address

TR-LANE requires Software Release 3.1(2) or later on the LightStream 100 switch and Cisco IOS Release 11.1(8) or later on the LightStream 1010.

For a complete description of the routers, switches, and interfaces, refer to your hardware documentation.

Addressing

On a LAN, packets are addressed by the MAC-layer address of the destination and source stations. To provide similar functionality for LANE, MAC-layer addressing must be supported. Every LANE client must have a MAC address. In addition, every LANE component (server, client, broadcast-and-unknown server, and configuration server) must have an ATM address that is different from that of all the other components.

All LANE clients on the same interface have the same, automatically assigned MAC address. That MAC address is also used as the end-system identifier (ESI) part of the ATM address, as explained in the next section. Although client MAC addresses are not unique, all ATM addresses are unique.

LANE ATM Addresses

A LANE ATM address has the same syntax as an NSAP, but it is not a network-level address. It consists of the following:

- A 13-byte prefix that includes the following fields defined by the ATM Forum:
  - AFI (Authority and Format Identifier) field (1 byte)
  - DCC (Data Country Code) or ICD (International Code Designator) field (2 bytes)
  - DFI field (Domain Specific Part Format Identifier) (1 byte)

    - Administrative Authority field (3 bytes)
    - Reserved field (2 bytes)
    - Routing Domain field (2 bytes)
    - Area field (2 bytes)
- A 6-byte end-system identifier (ESI)
- A 1-byte selector field

Cisco's Method of Automatically Assigning ATM Addresses

Cisco provides the following standard method of constructing and assigning ATM and MAC addresses for use in a LANE configuration server's database. A pool of MAC addresses is assigned to each ATM interface on the router. On the Cisco 7200 series routers, Cisco 7500 series routers, Cisco 4500 routers, and Cisco 4700 routers, the pool contains eight MAC addresses. For constructing ATM addresses, the following assignments are made to the LANE components:

- The prefix fields are the same for all LANE components in the router; the prefix indicates the identity of the switch. The prefix value must be configured on the switch.
- The ESI field value assigned to every *client* on the interface is the first of the pool of MAC addresses assigned to the interface.
- The ESI field value assigned to every *server* on the interface is the second of the pool of MAC addresses.
- The ESI field value assigned to the *broadcast-and-unknown server* on the interface is the third of the pool of MAC addresses.
- The ESI field value assigned to the *configuration server* is the fourth of the pool of MAC addresses.
- The selector field value is set to the subinterface number of the LANE component, except for the LANE configuration server, which has a selector field value of 0.

> Because the LANE components are defined on different subinterfaces of an ATM interface, the value of the selector field in an ATM address is different for each component. The result is a unique ATM address for each LANE component, even within the same router. For more information about assigning components to subinterfaces, see the "Rules for Assigning Components to Interfaces and Subinterfaces" section later in this chapter.

For example, if the MAC addresses assigned to an interface are 0800.200C.1000 through 0800.200C.1007, the ESI part of the ATM addresses is assigned to LANE components as follows:

- Any client gets the ESI 0800.200c.1000
- Any server gets the ESI 0800.200c.1001
- The broadcast-and-unknown server gets the ESI 0800.200c.1002
- The LANE configuration server gets the ESI 0800.200c.1003

Refer to the "Multiple Token Ring ELANs with Unrestricted Membership Example" and the "Multiple Token Ring ELANs with Restricted Membership Example" sections for examples using MAC address values as ESI field values in ATM addresses and for examples using subinterface numbers as selector field values in ATM addresses.

Using ATM Address Templates

ATM address templates can be used in many LANE commands that assign ATM addresses to LANE components (thus overriding automatically assigned ATM addresses) or that link client ATM addresses to emulated LANs. The use of templates can greatly simplify the use of these commands. The syntax of address templates, the use of address templates, and the use of wildcard characters within an address template for LANE are very similar to those for address templates of ISO CLNS.

NOTES

E.164-format ATM addresses do not support the use of LANE ATM address templates.

LANE ATM address templates can use two types of wildcards: an asterisk (*) to match any single character, and an ellipsis (...) to match any number of leading or trailing characters.

In LANE, a *prefix template* explicitly matches the prefix but uses wildcards for the ESI and selector fields. An *ESI template* explicitly matches the ESI field but uses wildcards for the prefix and selector. Table 8–1 indicates how the values of unspecified digits are determined when an ATM address template is used.

Table 8–1 *Values of Unspecified Digits in ATM Address Templates*

Unspecified Digits In	Value Is
Prefix (first 13 bytes)	Obtained from ATM switch via Interim Local Management Interface (ILMI)
ESI (next 6 bytes)	Filled with the slot MAC address* plus • 0—LANE client • 1—LANE server • 2—LANE broadcast-and-unknown server • 3—Configuration server
Selector field (last 1 byte)	Subinterface number, in the range 0 through 255.

* The lowest of the pool of MAC addresses assigned to the ATM interface plus a value that indicates the LANE component. For the Cisco 7200 series routers, Cisco 7500 series routers, Cisco 4500 routers, and Cisco 4700 routers, the pool has eight MAC addresses.

Rules for Assigning Components to Interfaces and Subinterfaces

The following rules apply to assigning LANE components to the major ATM interface and its subinterfaces in a given router:

- The LANE configuration server always runs on the major interface. (The assignment of any other component to the major interface is identical to assigning that component to the 0 subinterface.)
- The server and the client of the *same* emulated LAN can be configured on the same subinterface in a router.

- Clients of two *different* emulated LANs cannot be configured on the same subinterface in a router.
- Servers of two *different* emulated LANs cannot be configured on the same subinterface in a router.

LANE Configuration Task List

Before you begin to configure LANE, you must decide whether you want to set up one or multiple emulated LANs. If you set up multiple emulated LANs, you must also decide where the servers and clients will be located, and whether to restrict the clients that can belong to each emulated LAN. Bridged emulated LANs are configured just like any other LAN, in terms of commands and outputs. Once you have made those basic decisions, you can proceed to configure LANE.

To configure LANE, complete the tasks in the following sections:

- Create a LANE Plan and Worksheet
- Configuring the Prefix on the Switch
- Set Up the Signaling and ILMI PVCs
- Displaying LANE Default Addresses
- Entering the Configuration Server's ATM Address on the Cisco Switch
- Setting Up the Configuration Server's Database
- Enabling the Configuration Server
- Setting Up LANE Servers and Clients
- Configuring Fault-Tolerant Operation

Notes

While the preceding sections contain information about configuring SSRP fault tolerance, refer to the following section for detailed information about requirements and implementation considerations.

Once LANE is configured, you can monitor and maintain the components in the participating routers by completing the tasks in the "Monitoring and Maintaining the LANE Components" section later in this chapter.

For configuration examples, see the "LANE Configuration Examples" section at the end of this chapter.

Create a LANE Plan and Worksheet

It might help you to begin by drawing up a plan and a worksheet for your own LANE scenario, showing the following information and leaving space for noting the ATM address of each of the LANE components on each subinterface of each participating router:

- The router and interface where the LANE configuration server will be located.
- The router, interface, and subinterface where the LANE server and broadcast-and-unknown server for each emulated LAN will be located. There can be multiple servers for each emulated LAN for fault-tolerant operation.
- The routers, interfaces, and subinterfaces where the clients for each emulated LAN will be located.
- The name of the default emulated LAN (optional).
- The names of the emulated LANs that will have unrestricted membership.
- The names of the emulated LANs that will have restricted membership.

Notes

The last three items in the preceding list are very important; they determine how you set up each emulated LAN in the configuration server's database.

Configuring the Prefix on the Switch

Before you configure LANE components on any Cisco 7200 series router, Cisco 7500 series router, Cisco 4500 router, or Cisco 4700 router, you must configure the Cisco ATM switch with the ATM address prefix to be used by all LANE components in the switch cloud. On the Cisco switch, the ATM address prefix is called the node ID. Prefixes must be 26 digits long. If you provide fewer than 26 digits, zeros are added to the right of the specified value to fill it to 26 digits.

To set the ATM address prefix on the Cisco LightStream 1010, complete the following tasks on the Cisco switch, starting in global configuration mode:

Task	Command
Set the local node ID (prefix of the ATM address).	**atm address** {*atm-address* | *prefix...*}
Exit global configuration mode.	**exit**
Save the configuration values permanently.	**copy running-config startup-config**

To set the ATM address prefix on the Cisco LightStream 100, complete the following tasks on the Cisco switch:

Task	Command
Set the local node ID (prefix of the ATM address).	**set local** *name ip-address mask prefix*
Save the configuration values permanently.	**save**

On the Cisco switches, you can display the current prefix by using the **show network** command.

Notes

If you do not save the configured value permanently, it will be lost when the switch is reset or powered off.

SET UP THE SIGNALING AND ILMI PVCS

You must set up the signaling permanent virtual circuit (PVC) and the PVC that will communicate with the ILMI on the major ATM interface of any router that participates in LANE.

Complete this task only once for a major interface. You do not need to repeat this task on the same interface even though you might configure LANE servers and clients on several of its subinterfaces.

To set up these PVCs, complete the following steps, beginning in global configuration mode:

Task		Command
Step 1	Specify the major ATM interface and enter interface configuration mode.	
	• On the AIP for Cisco 7500 series routers; On the ATM port adapter for Cisco 7200 series routers.	**interface atm** *slot*/**0**
	• On the ATM port adapter for Cisco 7500 series routers.	**interface atm** *slot*/*port-adapter*/**0**
	• On the NPM for Cisco 4500 and Cisco 4700 routers.	**interface atm** *number*
Step 2	Set up the signaling PVC that sets up and tears down switched virtual circuits (SVCs); the *vpi* and *vci* values are usually set to 0 and 5, respectively.	**atm pvc** *vcd vpi vci* **qsaal**
Step 3	Set up a PVC to communicate with the ILMI; the *vpi* and *vci* values are usually set to 0 and 16, respectively.	**atm pvc** *vcd vpi vci* **ilmi**

DISPLAYING LANE DEFAULT ADDRESSES

You can display the LANE default addresses to make configuration easier. Complete this task for each router that participates in LANE. This command displays default addresses for all ATM interfaces present on the router. Write down the displayed addresses on your worksheet.

To display the default LANE addresses, perform the following task in global configuration mode:

Task	Command
Display the LANE default addresses.	**show lane default-atm-addresses**

Entering the Configuration Server's ATM Address on the Cisco Switch

You must enter the configuration server's ATM address into the Cisco LightStream 100 or Cisco LightStream 1010 ATM switch and save it permanently so that the value is not lost when the switch is reset or powered off. You also must specify the full 40-digit ATM address. Use the addresses on your worksheet that you obtained from the previous task.

If you are configuring Simple Server Redundancy Protocol (SSRP), enter the multiple LANE configuration server addresses into the end ATM switches. The switches are used as central locations for the list of LANE configuration server addresses. LANE components connected to the switches obtain the global list of LANE configuration server addresses from the switches.

Depending on which type of switch you are using, perform one of the following tasks:

- Entering the ATM Address(es) on the Cisco LightStream 1010 ATM Switch
- Entering the ATM Address(es) on the Cisco LightStream 100 ATM Switch

Entering the ATM Address(es) on the Cisco LightStream 1010 ATM Switch

On the Cisco LightStream 1010 ATM switch, the configuration server address can be specified for a port or for the entire switch. To enter the configuration server addresses on the Cisco LightStream 1010 ATM switch for the entire switch, perform the following tasks beginning in global configuration mode.

To enter the configuration server addresses on the Cisco LightStream 1010 ATM switch per port, perform the following tasks beginning in interface configuration mode:

Task		Command
Step 1	Specify the LANE configuration server's ATM address for a port. If you are configuring SSRP, include the ATM addresses of all the LANE configuration servers.	**atm lecs-address** *lecsaddress* [*sequence #*]*
Step 2	Exit interface configuration mode.	**Ctrl-Z**
Step 3	Save the configuration value permanently.	**copy running-config startup-config**

* Refer to the *LightStream 1010 ATM Switch Command Reference* for further information about this command.

Entering the ATM Address(es) on the Cisco LightStream 100 ATM Switch

To enter the configuration server's ATM address into the Cisco LightStream 100 ATM switch and save it permanently, perform the following tasks in privileged EXEC mode:

Task		Command
Step 1	Specify the LANE configuration server's ATM address. If you are configuring SSRP, repeat this command for each configuration server address. The *index* value determines the priority. The highest priority is 0. There can be a maximum of 4 LANE configuration servers.	**set configserver** *index atm-address*
Step 2	Save the configuration value permanently.	**save**

Setting Up the Configuration Server's Database

The configuration server's database contains information about each emulated LAN, including the ATM addresses of the LANE servers. You can specify one default emulated LAN in the database. The LANE configuration server will assign any client that does not request a specific emulated LAN to the default emulated LAN.

Emulated LANs are either restricted or unrestricted. The configuration server will assign a client to an unrestricted emulated LAN if the client specifies that particular emulated LAN in its configuration. However, the configuration server will only assign a client to a restricted emulated LAN if the client is specified in the configuration server's database as belonging to that emulated LAN. The default emulated LAN must have unrestricted membership.

If you are configuring fault tolerance, you can have any number of servers per emulated LAN. Priority is determined by entry order; the first entry has the highest priority, unless you override it with the index option.

To set up the database, complete the tasks in the following sections as appropriate for your emulated LAN plan and scenario:

- Setting Up the Database for the Default Emulated LAN Only
- Setting Up the Database for Unrestricted-Membership Emulated LANs
- Setting Up the Database for Restricted-Membership LANs

Setting Up the Database for the Default Emulated LAN Only

When you configure a router as the configuration server for one default emulated LAN, you provide a name for the database, the ATM address of the LANE server for the emulated LAN and a default name for the emulated LAN. In addition, you indicate that the configuration server's ATM address is to be computed automatically.

When you configure a database with only a default unrestricted emulated LAN, you do not have to specify where the LANE clients are located. That is, when you set up the configuration server's database for a single default emulated LAN, you do not have to provide any database entries that link the ATM addresses of any clients with the emulated LAN name. All of the clients are assigned to the default emulated LAN.

To set up the configuration server for the default emulated LAN, complete the following tasks beginning in global configuration mode:

Task		Commands
Step 1	Create a named database for the LANE configuration server.	**lane database** *database-name*
Step 2	In the configuration database, bind the name of the emulated LAN to the ATM address of the LANE server. If you are configuring SSRP, repeat this step for each additional server for the same emulated LAN. The index determines the priority. The highest priority is 0.	**name** *elan-name* **server-atm-address** *atm-address* [**index** *number*]
Step 3	If you are configuring a Token Ring emulated LAN, assign a segment number to the emulated Token Ring LAN in the configuration database.	**name** *elan-name* **local-seg-id** *segment-number*
Step 4	In the configuration database, provide a default name for the emulated LAN.	**default-name** *elan-name*
Step 5	Exit from database configuration mode and return to global configuration mode.	**exit**

In Step 2, enter the ATM address of the server for the specified emulated LAN, as noted in your worksheet and obtained in the "Displaying LANE Default Addresses" section earlier in this chapter.

You can have any number of servers per emulated LAN for fault tolerance. Priority is determined by entry order. The first entry has the highest priority unless you override it with the index option. If you are setting up only a default emulated LAN, the *elan-name* value in Steps 2 and 3 is the same as the default emulated LAN name you provide in Step 4.

To set up fault-tolerant operation, see the "Configuring Fault-Tolerant Operation" section later in this chapter.

Setting Up the Database for Unrestricted-Membership Emulated LANs

When you set up a database for unrestricted emulated LANs, you create database entries that link the name of each emulated LAN to the ATM address of its server; however, you may choose not to specify where the LANE clients are located. That is, when you set up the configuration server's database, you do not have to provide any database entries that link the ATM addresses or MAC addresses of any clients with the emulated LAN name. The configuration server will assign the clients to the emulated LANs specified in the client's configurations.

To configure a router as the configuration server for multiple emulated LANs with unrestricted membership, perform the following tasks beginning in global configuration mode:

Task		Command
Step 1	Create a named database for the LANE configuration server.	**lane database** *database-name*
Step 2	In the configuration database, bind the name of the first emulated LAN to the ATM address of the LANE server for that emulated LAN. If you are configuring SSRP, repeat this step with the same emulated LAN name but with different server ATM addresses for each additional server for the same emulated LAN. The index determines the priority. The highest priority is 0.	**name** *elan-name1* **server-atm-address** *atm-address* [**index** *number*]
Step 3	In the configuration database, bind the name of the second emulated LAN to the ATM address of the LANE server. If you are configuring SSRP, repeat this step with the same emulated LAN name but with different server ATM addresses for each additional server for the same emulated LAN. The index determines the priority. The highest priority is 0. Repeat this step, providing a different emulated LAN name and ATM address for each additional emulated LAN in this switch cloud.	**name** *elan-name2* **server-atm-address** *atm-address* [**index** *number*]

Task		Command
Step 4	For a Token Ring emulated LAN, assign a segment number to the first emulated Token Ring LAN in the configuration database.	**name** *elan-name1* **local-seg-id** *segment-number*
Step 5	For Token Ring emulated LANs, assign a segment number to the second emulated Token Ring LAN in the configuration database. Repeat this step, providing a different emulated LAN name and segment number for each additional source-route bridged emulated LAN in this switch cloud.	**name** *elan-name2* **local-seg-id** *segment-number*
Step 6	(Optional) Specify a default emulated LAN for LANE clients not explicitly bound to an emulated LAN.	**default-name** *elan-name1*
Step 7	Exit from database configuration mode and return to global configuration mode.	**exit**

In the preceding steps, enter the ATM address of the server for the specified emulated LAN, as noted in your worksheet and obtained in the "Displaying LANE Default Addresses" section. To set up fault-tolerant operation, see the "Configuring Fault-Tolerant Operation" section later in this chapter.

Setting Up the Database for Restricted-Membership LANs

When you set up the database for restricted-membership emulated LANs, you create database entries that link the name of each emulated LAN to the ATM address of its server; however, you must also specify where the LANE clients are located. That is, for each restricted-membership emulated LAN, you provide a database entry that explicitly links the ATM address or MAC address of each client of that emulated LAN with the name of that emulated LAN.

The client database entries specify which clients are allowed to join the emulated LAN. When a client requests to join an emulated LAN, the configuration server consults its database and then assigns the client to the emulated LAN specified in the configuration server's database.

When clients for the same restricted-membership emulated LAN are located in multiple routers, each client's ATM address or MAC address must be linked explicitly with the name of the emulated LAN. As a result, you must configure as many client entries (at Steps 6 and 7 in the following procedure) as you have clients for emulated LANs in all the routers. Each client will have a different ATM address in the database entries.

To set up the configuration server for emulated LANs with restricted membership, perform the following tasks beginning in global configuration mode:

Task		Command
Step 1	Create a named database for the LANE configuration server.	**lane database** *database-name*
Step 2	In the configuration database, bind the name of the first emulated LAN to the ATM address of the LANE server for that emulated LAN. If you are configuring SSRP, repeat this step with the same emulated LAN name but with different server ATM addresses for each additional server for the same emulated LAN. The index determines the priority. The highest priority is 0.	**name** *elan-name1* **server-atm-address** *atm-address* **restricted** [**index** *number*]
Step 3	In the configuration database, bind the name of the second emulated LAN to the ATM address of the LANE server. If you are configuring SSRP, repeat this step with the same emulated LAN name but with different server ATM addresses for each additional server for the same emulated LAN. The index determines the priority. The highest priority is 0. Repeat this step, providing a different name and a different ATM address for each additional emulated LAN.	**name** *elan-name2* **server-atm-address** *atm-address* **restricted** [**index** *number*]

Step 4	For a Token Ring emulated LAN, assign a segment number to the first emulated Token Ring LAN in the configuration database.	**name** *elan-name1* **local-seg-id** *segment-number*
Step 5	If you are configuring Token Ring emulated LANs, assign a segment number to the second emulated Token Ring LAN in the configuration database. Repeat this step, providing a different emulated LAN name and segment number for each additional source-route bridged emulated LAN in this switch cloud.	**name** *elan-name2* **local-seg-id** *segment-number*
Step 6	Add a database entry associating a specific client's ATM address with the first restricted-membership emulated LAN. Repeat this step for each of the clients of the first restricted-membership emulated LAN.	**client-atm-address** *atm-address-template* **name** *elan-name1*
Step 7	Add a database entry associating a specific client's ATM address with the second restricted-membership emulated LAN. Repeat this step for each of the clients of the second restricted-membership emulated LAN. Repeat this step, providing a different name and a different list of client ATM address, for each additional emulated LAN.	**client-atm-address** *atm-address-template* **name** *elan-name2*
Step 8	Exit from database configuration mode and return to global configuration mode.	**exit**

Enabling the Configuration Server

After you create the database, you can enable the configuration server on the selected ATM interface and router by completing the following tasks, beginning in global configuration mode:

Task		Command
Step 1	If you are not currently configuring the interface, specify the major ATM interface where the configuration server is located.	
	• On the AIP for Cisco 7500 series routers; On the ATM port adapter for Cisco 7200 series routers.	**interface atm** *slot*/0[.*subinterface-number*]
	• On the ATM port adapter for Cisco 7500 series routers.	**interface atm** *slot*/*port-adapter*/0 [.*subinterface-number*]
	• On the NPM for Cisco 4500 and Cisco 4700 routers.	**interface atm** *number*[.*subinterface-number*]
Step 2	Link the configuration server's database name to the specified major interface, and enable the configuration server.	**lane config database** *database-name*
Step 3	Specify how the LECS's ATM address will be computed. You may opt to choose one of the following scenarios:	
	• The LECS will participate in SSRP and the address is computed by the automatic method.	**lane config auto-config-atm-address**
	• The LECS will participate in SSRP, and the address is computed by the automatic method. If the LECS is the master, the fixed address is also used.	**lane config auto-config-atm-address** **lane config fixed-config-atm-address**
	• The LECS will not participate in SSRP, the LECS is the master, and only the well-known address is used.	**lane config fixed-config-atm-address**
	• The LECS will participate in SSRP and the address is computed using an explicit, 20-byte ATM address.	**lane config config-atm-address** *atm-address-template*
Step 4	Exit interface configuration mode.	**exit**
Step 5	Return to EXEC mode.	**Ctrl-Z**
Step 6	Save the configuration.	**copy running-config startup-config**

Setting Up LANE Servers and Clients

For each router that will participate in LANE, set up the necessary servers and clients for each emulated LAN. Then display and record the server and client ATM addresses. Be sure to keep track of the router interface where the LANE configuration server will eventually be located.

You can set up servers for more than one emulated LAN on different subinterfaces or on the same interface of a router, or you can place the servers on different routers. When you set up a server and broadcast-and-unknown server on a router, you can combine them with a client on the same subinterface, a client on a different subinterface, or no client at all on the router. Where you put the clients is important because any router with clients for multiple emulated LANs can route frames between those emulated LANs.

Depending on where your clients and servers are located, perform one of the following tasks for each LANE subinterface.

- Setting Up the Server, Broadcast-and-Unknown Server, and a Client on a Subinterface
- Setting Up Only a Client on a Subinterface

Setting Up the Server, Broadcast-and-Unknown Server, and a Client on a Subinterface

To set up the server, broadcast-and-unknown server, and (optionally) clients for an emulated LAN, perform the following tasks beginning in global configuration mode:

Task		Command
Step 1	Specify the subinterface for the emulated LAN on this router.	
	• On the AIP for Cisco 7500 series routers; On the ATM port adapter for Cisco 7200 series routers.	**interface atm** *slot*/0.*subinterface-number*
	• On the ATM port adapter for Cisco 7500 series routers.	**interface atm** *slot*/*port-adapter*/0.*subinterface-number*
	• On the NPM for Cisco 4500 and Cisco 4700 routers.	**interface atm** *number*.*subinterface-number*

Task		Command
Step 2	Enable a LANE server and a LANE broadcast-and-unknown server for the emulated LAN.	**lane server-bus** {**ethernet** | **tokenring**} *elan-name*
Step 3	(Optional) Enable a LANE client for the emulated LAN.	**lane client** {**ethernet** | **tokenring**} [*elan-name*]
Step 4	Provide a protocol address for the client.	**ip** *address mask*
Step 5	Return to EXEC mode.	**Ctrl-Z**
Step 6	Save the configuration.	**copy running-config startup-config**

If the emulated LAN in Step 3 is intended to have *restricted membership*, consider carefully whether you want to specify its name here. You will specify the name in the LANE configuration server's database when it is set up. However, if you link the client to an emulated LAN in this step, and through some mistake it does not match the database entry linking the client to an emulated LAN, this client will not be allowed to join this emulated LAN or any other.

If you do decide to include the name of the emulated LAN linked to the client in Step 3 and later want to associate that client with a different emulated LAN, make the change in the configuration server's database before you make the change for the client on this subinterface.

Each emulated LAN is a separate subnetwork. In Step 4 make sure that the clients of the same emulated LAN are assigned protocol addresses on the same subnetwork and that clients of different emulated LANs are assigned protocol addresses on different subnetworks.

Setting Up Only a Client on a Subinterface

On any given router, you can set up one client for one emulated LAN or multiple clients for multiple emulated LANs. You can set up a client for a given emulated LAN on any routers you choose to participate in that emulated LAN. Any router with clients for multiple emulated LANs can route packets between those emulated LANs.

You must first set up the signaling and ILMI PVCs on the major ATM interface, as described earlier in the "Set Up the Signaling and ILMI PVCs" section before you set up the client.

To set up only a client for an emulated LANs, perform the following tasks beginning in interface configuration mode:

Task		Command
Step 1	Specify the subinterface for the emulated LAN on this router.	
	• On the AIP for Cisco 7500 series routers; on the ATM port adapter for Cisco 7200 series routers.	**interface atm** *slot*/0.*subinterface-number*
	• On the ATM port adapter for Cisco 7500 series routers.	**interface atm** *slot*/*port-adapter*/0.*subinterface-number*
	• On the NPM for Cisco 4500 and Cisco 4700 routers.	**interface atm** *number.subinterface-number*
Step 2	Provide a protocol address for the client on this subinterface.	**ip** *address mask*
Step 3	Enable a LANE client for the emulated LAN.	**lane client** {**ethernet** | **tokenring**} [*elan-name*]
Step 4	Return to EXEC mode.	**Ctrl-Z**
Step 5	Save the configuration.	**copy running-config startup-config**

Each emulated LAN is a separate subnetwork. In Step 2, make sure that the clients of the same emulated LAN are assigned protocol addresses on the same subnetwork and that clients of different emulated LANs are assigned protocol addresses on different subnetworks.

Configuring Fault-Tolerant Operation

The LANE simple server redundancy feature creates fault tolerance using standard LANE protocols and mechanisms. If a failure occurs on the LANE configuration server or on the LANE server/broadcast-and-unknown server, the emulated LAN can continue to operate using the services of a backup LANE server. This protocol is called the Simple Server Redundancy Protocol (SSRP).

This section describes how to configure simple server redundancy for fault tolerance on an emulated LAN.

Notes

This server redundancy does not overcome other points of failure beyond the router ports: Additional redundancy on the LAN side or in the ATM switch cloud are not a part of the LANE simple server redundancy feature.

Simple Server Redundancy Requirements

For simple LANE service replication or fault tolerance to work, the ATM switch must support multiple LANE server addresses. This mechanism is specified in the LANE standard. The LANE servers establish and maintain a standard control circuit that enables the server redundancy to operate.

LANE simple server redundancy is supported on Cisco IOS Release 11.2 and later software. Older LANE configuration files continue to work with this new software.

This redundancy feature works only with Cisco LANE configuration servers and LANE server/broadcast-and-unknown server combinations. Third-party LANE Clients can be used with the SSRP, but third-party configuration servers, LANE servers, and broadcast-and-unknown servers do not support SSRP.

For server redundancy to work correctly, the following protocol must be met:

- All the ATM switches must have identical lists of the global LANE configuration server addresses in the identical priority order.

- The operating LANE configuration servers must use exactly the same configuration database. Load the configuration table data using the **copy** {**rcp** | **tftp**} **running-config** command. This method minimizes errors and enables the database to be maintained centrally in one place.

The LANE protocol does not specify where any of the emulated LAN server entities should be located, but for the purpose of reliability and performance, Cisco implements these server components on its routers.

Redundant Configuration Servers

To enable redundant LANE configuration servers, enter the multiple LANE configuration server addresses into the end ATM switches. LANE components can obtain the list of LANE configuration server addresses from the ATM switches through the Interim Local Management Interface (ILMI).

Refer to the "Entering the Configuration Server's ATM Address on the Cisco Switch" section earlier in this chapter for more details.

Redundant Servers and Broadcast-and-Unknown Servers

The LANE configuration server turns on server/broadcast-and-unknown server redundancy by adjusting its database to accommodate multiple server ATM addresses for a particular emulated LAN. The additional servers serve as backup servers for that emulated LAN.

To activate the feature, you add an entry for the hierarchical list of servers that will support the given emulated LAN. All database modifications for the emulated LAN must be identical on all LANE configuration servers. Refer to the "Setting Up the Configuration Server's Database" section earlier in this chapter for more details.

Implementation Considerations

Following are considerations for implementation:

- Up to 16 LANE configuration server addresses can be handled by the LANE subsystem. The LightStream 100 allows a maximum of 4 LANE configuration server addresses.
- There is no limit on the number of LANE servers that can be defined per emulated LAN.
- When a LANE configuration server switchover occurs, no previously joined clients are affected.
- When a LANE server/broadcast-and-unknown server switches over, momentary loss of clients occurs until they are all transferred to the new LANE server/broadcast-and-unknown server.
- LANE configuration servers come up as masters until a higher-level LANE configuration server tells them otherwise. This is automatic and cannot be changed.
- If a higher-priority LANE server comes online, it bumps the current LANE server off on the same emulated LAN. Therefore, there may be some flapping of clients from one LANE server to another after a powerup, depending on the order of the LANE servers coming up. Flapping should settle after the *last* highest-priority LANE server comes up.
- If none of the specified LANE servers are up or connected to the master LANE configuration server and more than one LANE server is defined for an emulated LAN, a configuration request for that specific emulated LAN is rejected by the LANE configuration server.
- Changes made to the list of LANE configuration server addresses on ATM switches may take up to a minute to propagate through the network. Changes made to the configuration database regarding LANE server addresses take effect almost immediately.

- If none of the designated LANE configuration servers are operational or reachable, the ATM Forum-defined well-known LANE configuration server address is used.
- You can override the LANE configuration server address on any subinterface by using the following commands:
  - **lane auto-config-atm-address**
  - **lane fixed-config-atm-address**
  - **lane config-atm-address**

CAUTION

When an override like this is performed, fault-tolerant operation cannot be guaranteed. To avoid affecting the fault-tolerant operation, do not override any LANE configuration server, LANE server or broadcast-and-unknown server addresses.

- If an underlying ATM network failure occurs, there may be multiple master LANE configuration servers and multiple active LANE servers for the same emulated LAN. This situation creates a "partitioned" network. The clients continue to operate normally, but transmission between different partitions of the network is not possible. When the network break is repaired, the system recovers.
- When the LECS is already up and running, and you use the **lane config fixed-config-atm-address** command to configure the well-known LECS address, please be aware of the following scenarios:
  - If you configure the LECS with only the well-known address, the LECS will not participate in the SSRP, act as a "standalone" master, and only listen on the well-known LECS address. This scenario is ideal if you want a "standalone" LECS that does not participate in SSRP, and you would like to listen to only the well-known address.
  - If only the well-known address is already assigned, and you assign at least one other address to the LECS, (additional addresses are assigned using the **lane config auto-config-atm-address** command or the **lane config config-atm-address command**) the LECS will participate in the SSRP and act as the master or slave

based on the normal SSRP rules. This scenario is ideal if you would like the LECS to participate in SSRP, and you would like to make the master LECS listen on the well-known address.

- If the LECS is participating in SSRP, has more than one address (one of which is the well-known address), and all the addresses but the well-known address is removed, the LECS will declare itself the master and stop participating in SSRP completely.
- If the LECS is operating as an SSRP slave, and it has the well-known address configured, it will not listen on the well-known address unless it becomes the master.
- If you want the LECS to assume the well-known address only when it becomes the master, configure the LECS with the well-known address and at least one other address.

Monitoring and Maintaining the LANE Components

After configuring LANE components on an interface or any of its subinterfaces, on a specified subinterface, or on an emulated LAN, you can display their status. To show LANE information, perform the following tasks in EXEC mode:

Task	Command
Display the global and per-virtual channel connection LANE information for all the LANE components and emulated LANs configured on an interface or any of its subinterfaces.	
• On the AIP for Cisco 7500 series routers; on the ATM port adapter for Cisco 7200 series routers.	**show lane** [**interface atm** *slot*/0 [.*subinterface-number*] | **name** *elan-name*] [**brief**]
• On the ATM port adapter for Cisco 7500 series routers.	**show lane** [**interface atm** *slot*/*port-adapter*/0 [.*subinterface-number*] | **name** *elan-name*] [**brief**]
• On the NPM for Cisco 4500 and Cisco 4700 routers.	**show lane** [**interface atm** *number* [.*subinterface-number*] | **name** *elan-name*] [**brief**]

Task	Command
Display the global and per-VCC LANE information for the broadcast-and-unknown server configured on any subinterface or emulated LAN.	
• On the AIP for Cisco 7500 series routers; on the ATM port adapter for Cisco 7200 series routers.	**show lane bus [interface atm** *slot*/**0** [.*subinterface-number*] | **name** *elan-name*] [**brief**]
• On the ATM port adapter for Cisco 7500 series routers.	**show lane bus [interface atm** *slot*/*port-adapter*/ **0** [.*subinterface-number*] | **name** *elan-name*] [**brief**]
• On the NPM for Cisco 4500 and Cisco 4700 routers.	**show lane bus [interface atm** *number* [.*subinterface-number*] | **name** *elan-name*] [**brief**]
Display the global and per-VCC LANE information for all LANE clients configured on any subinterface or emulated LAN.	
• On the AIP for Cisco 7500 series routers; on the ATM port adapter for Cisco 7200 series routers.	**show lane client [interface atm** *slot*/**0** [.*subinterface-number*] | **name** *elan-name*] [**brief**]
• On the ATM port adapter for Cisco 7500 series routers.	**show lane client [interface atm** *slot*/*port-adapter*/**0** [.*subinterface-number*] | **name** *elan-name*] [**brief**]
• On the NPM for Cisco 4500 and Cisco 4700 routers.	**show lane client [interface atm** *number* [.*subinterface-number*] | **name** *elan-name*] [**brief**]
Display the global and per-VCC LANE information for the configuration server configured on any interface.	
• On the AIP for Cisco 7500 series routers; on the ATM port adapter for Cisco 7200 series routers.	**show lane config [interface atm** *slot*/**0**]
• On the ATM port adapter for Cisco 7500 series routers.	**show lane config [interface atm** *slot*/*port-adapter*/**0**]
• On the NPM for Cisco 4500 and Cisco 4700 routers.	**show lane config [interface atm** *number*]

Task	Command
Display the LANE configuration server's database.	**show lane database** [*database-name*]
Display the automatically assigned ATM address of each LANE component in a router or on a specified interface or subinterface.	
• On the AIP for Cisco 7500 series routers; on the ATM port adapter for Cisco 7200 series routers.	**show lane default-atm-addresses** [**interface atm** *slot*/0.*subinterface-number*]
• On the ATM port adapter for Cisco 7500 series routers.	**show lane default-atm-addresses** [**interface atm** *slot*/*port-adapter*/0.*subinterface-number*]
• On the NPM for Cisco 4500 and Cisco 4700 routers.	**show lane default-atm-addresses** [**interface atm** *number.subinterface-number*]
Display the LANE ARP table of the LANE client configured on the specified subinterface or emulated LAN.	
• On the AIP for Cisco 7500 series routers; on the ATM port adapter for Cisco 7200 series routers.	**show lane le-arp** [**interface atm** *slot*/0[.*subinterface-number*] | **name** *elan-name*]
• On the ATM port adapter for Cisco 7500 series routers.	**show lane le-arp** [**interface atm** *slot*/*port-adapter*/0[.*subinterface-number*] | **name** *elan-name*]
• On the NPM for Cisco 4500 and Cisco 4700 routers.	**show lane le-arp** [**interface atm** *number* [.*subinterface-number*] | **name** *elan-name*]

Task	Command
Display the global and per-VCC LANE information for the LANE server configured on a specified subinterface or emulated LAN.	
• On the AIP for Cisco 7500 series routers; on the ATM port adapter for Cisco 7200 series routers.	**show lane server** [**interface atm** *slot*/0 [.*subinterface-number*] | **name** *elan-name*] [**brief**]
• On the ATM port adapter for Cisco 7500 series routers.	**show lane server** [**interface atm** *slot*/*port-adapter*/0[.*subinterface-number*] | **name** *elan-name*] [**brief**]
• On the NPM for Cisco 4500 and Cisco 4700 routers.	**show lane server** [**interface atm** *number* [.*subinterface-number*] | **name** *elan-name*] [**brief**]

LANE Configuration Examples

The examples in the following sections illustrate how to configure LANE for the following cases:

- Default Configuration for a Single Ethernet Emulated LAN Example
- Default Configuration for a Single Ethernet Emulated LAN with a Backup LANE Configuration Server and LANE Server Example
- Multiple Token Ring ELANs with Unrestricted Membership Example
- Multiple Token Ring ELANs with Restricted Membership Example
- TR-LANE with 2-Port Source-Route Bridging Example
- TR-LANE with Multiport Source-Route Bridging Example
- Routing Between Token Ring and Ethernet Emulated LANs Example

All examples use the automatic ATM address assignment method described in the "Cisco's Method of Automatically Assigning ATM Addresses" section earlier in this chapter. These examples show the LANE configurations, not the process of determining the ATM addresses and entering them.

Default Configuration for a Single Ethernet Emulated LAN Example

The following example configures four Cisco 7500 series routers for one Ethernet emulated LAN. Router 1 contains the configuration server, the server, the broadcast-and-unknown server, and a client. The remaining routers each contain a client for the emulated LAN. This example accepts all default settings that are provided. For example, it does not explicitly set ATM addresses for the different LANE components that are colocated on the router. Membership in this LAN is not restricted.

Router 1

```
lane database example1
 name eng server-atm-address 39.000001415555121101020304.0800.200c.1001.01
 default-name eng
interface atm 1/0
 atm pvc 1 0 5 qsaal
 atm pvc 2 0 16 ilmi
 lane config auto-config-atm-address
 lane config database example1
interface atm 1/0.1
 ip address 172.16.0.1 255.255.255.0
 lane server-bus ethernet eng
 lane client ethernet
```

Router 2

```
interface atm 1/0
 atm pvc 1 0 5 qsaal
 atm pvc 2 0 16 ilmi
interface atm 1/0.1
 ip address 172.16.0.3 255.255.255.0
 lane client ethernet
```

Router 3

```
interface atm 2/0
 atm pvc 1 0 5 qsaal
 atm pvc 2 0 16 ilmi
interface atm 2/0.1
 ip address 172.16.0.4 255.255.255.0
 lane client ethernet
```

Router 4

```
interface atm 1/0
 atm pvc 1 0 5 qsaal
```

```
 atm pvc 2 0 16 ilmi
interface atm 1/0.3
 ip address 172.16.0.5 255.255.255.0
 lane client ethernet
```

Default Configuration for a Single Ethernet Emulated LAN with a Backup LANE Configuration Server and LANE Server Example

This example configures four Cisco 7500 series routers for one emulated LAN with fault tolerance. Router 1 contains the configuration server, the server, the broadcast-and-unknown server, and a client. Router 2 contains the backup LANE configuration server and the backup LANE server for this emulated LAN and another client. Routers 3 and 4 contain clients only. This example accepts all default settings that are provided. For example, it does not explicitly set ATM addresses for the various LANE components located on the router. Membership in this LAN is not restricted.

Router 1

```
lane database example1
 name eng server-atm-address 39.000001415555121101020304.0800.200c.1001.01
 name eng server-atm-address 39.000001415555121101020304.0612.200c 2001.01
 default-name eng
interface atm 1/0
 atm pvc 1 0 5 qsaal
 atm pvc 2 0 16 ilmi
 lane config auto-config-atm-address
 lane config database example1
interface atm 1/0.1
 ip address 172.16.0.1 255.255.255.0
 lane server-bus ethernet eng
 lane client ethernet
```

Router 2

```
lane database example1_backup
 name eng server-atm-address 39.000001415555121101020304.0800.200c.1001.01
 name eng server-atm-address 39.000001415555121101020304.0612.200c 2001.01 (backup LES)
 default-name eng
interface atm 1/0
 atm pvc 1 0 5 qsaal
 atm pvc 2 0 16 ilmi
 lane config auto-config-atm-address
 lane config database example1_backup
interface atm 1/0.1
```

```
 ip address 172.16.0.3 255.255.255.0
 lane server-bus ethernet eng
 lane client ethernet
```

Router 3

```
interface atm 2/0
 atm pvc 1 0 5 qsaal
 atm pvc 2 0 16 ilmi
interface atm 2/0.1
 ip address 172.16.0.4 255.255.255.0
 lane client ethernet
```

Router 4

```
interface atm 1/0
 atm pvc 1 0 5 qsaal
 atm pvc 2 0 16 ilmi
interface atm 1/0.3
 ip address 172.16.0.5 255.255.255.0
 lane client ethernet
```

Multiple Token Ring ELANs with Unrestricted Membership Example

The following example configures four Cisco 7500 series routers for three emulated LANs for Engineering, Manufacturing, and Marketing, as illustrated in Figure 8–1. This example does not restrict membership in the emulated LANs.

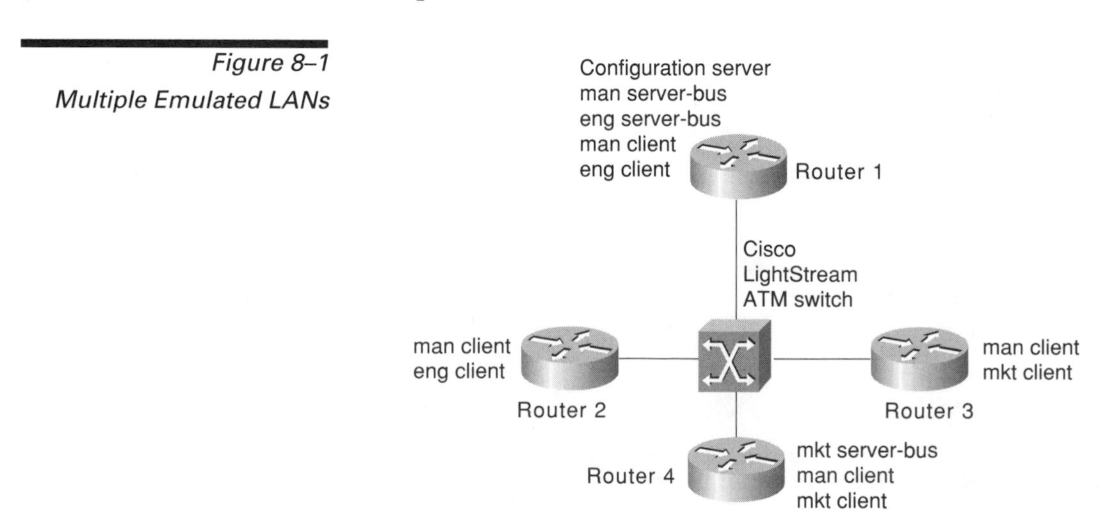

Figure 8–1
Multiple Emulated LANs

In this example, Router 1 has the following LANE components:

- The LANE configuration server (there is one configuration server for this group of emulated LANs)
- The LANE server and broadcast-and-unknown server for the emulated LAN for Manufacturing (*man*)
- The LANE server and broadcast-and-unknown server for the emulated LAN for Engineering (*eng*)
- A LANE client for the emulated LAN for Manufacturing (*man*)
- A LANE client for the emulated LAN for Engineering (*eng*)

Router 2 has the following LANE components:

- A LANE client for the emulated LAN for Manufacturing (*man*)
- A LANE client for the emulated LAN for Engineering (*eng*)

Router 3 has the following LANE components:

- A LANE client for the emulated LAN for Manufacturing (*man*)
- A LANE client for the emulated LAN for Marketing (*mkt*)

Router 4 has the following LANE components:

- The LANE server and broadcast-and-unknown server for the emulated LAN for Marketing (*mkt*)
- A LANE client for the emulated LAN for Manufacturing (*man*)
- A LANE client for the emulated LAN for Marketing (*mkt*)

For the purposes of this example, the four routers are assigned ATM address prefixes and end system identifiers (ESIs) as shown in Table 8–2 (the ESI part of the ATM address is derived from the first MAC address of the AIP shown in the example).

Table 8–2 *ATM Prefixes for TR-LANE Example*

Router	ATM Address Prefix	ESI Base
Router 1	39.000001415555121101020304	0800.200c.1000
Router 2	39.000001415555121101020304	0800.200c.2000
Router 3	39.000001415555121101020304	0800.200c.3000
Router 4	39.000001415555121101020304	0800.200c.4000

Router 1

Router 1 has the configuration server and its database, the server and broadcast-and-unknown server for the Manufacturing emulated LAN, the server and broadcast-and-unknown server for the Engineering emulated LAN, a client for Manufacturing, and a client for Engineering. Router 1 is configured as shown in this example:

```
!The following lines name and configure the configuration server's database.
lane database example2
 name eng server-atm-address 39.000001415555121101020304.0800.200c.1001.02
 name eng local-seg-id 1000
 name man server-atm-address 39.000001415555121101020304.0800.200c.1001.01
 name man local-seg-id 2000
 name mkt server-atm-address 39.000001415555121101020304.0800.200c.4001.01
 name mkt local-seg-id 3000
 default-name man
!
! The following lines bring up the configuration server and associate
! it with a database name.
interface atm 1/0
 atm pvc 1 0 5 qsaal
 atm pvc 2 0 16 ilmi
 lane config auto-config-atm-address
 lane config database example2
!
! The following lines configure the "man" server, broadcast-and-unknown server,
! and the client on atm subinterface 1/0.1. The client is assigned to the default
! emulated lan.
interface atm 1/0.1
 ip address 172.16.0.1 255.255.255.0
 lane server-bus tokenring man
 lane client tokenring man
!
! The following lines configure the "eng" server, broadcast-and-unknown server,
! and the client on atm subinterface 1/0.2. The client is assigned to the
! engineering emulated lan. Each emulated LAN is a different subnetwork, so the "eng"
```

```
! client has an IP address on a different subnetwork than the "man" client.
interface atm 1/0.2
 ip address 172.16.1.1 255.255.255.0
 lane server-bus tokenring eng
 lane client tokenring eng
```

Router 2

Router 2 is configured for a client of the Manufacturing emulated LAN and a client of the Engineering emulated LAN. Because the default emulated LAN name is *man*, the first client is linked to that emulated LAN name by default. Router 2 is configured as shown in the following code:

```
interface atm 1/0
 atm pvc 1 0 5 qsaal
 atm pvc 2 0 16 ilmi
interface atm 1/0.1
 ip address 172.16.0.2 255.255.255.0
 lane client tokenring
interface atm 1/0.2
 ip address 172.16.1.2 255.255.255.0
 lane client tokenring eng
```

Router 3

Router 3 is configured for a client of the Manufacturing emulated LAN and a client of the Marketing emulated LAN. Because the default emulated LAN name is *man*, the first client is linked to that emulated LAN name by default. Router 3 is configured as shown in the following code:

```
interface atm 2/0
 atm pvc 1 0 5 qsaal
 atm pvc 2 0 16 ilmi
interface atm 2/0.1
 ip address 172.16.0.3 255.255.255.0
 lane client tokenring
interface atm 2/0.2
 ip address 172.16.2.3 255.255.255.0
 lane client tokenring mkt
```

Router 4

Router 4 has the server and broadcast-and-unknown server for the Marketing emulated LAN, a client for Marketing, and a client for Manufacturing. Because the default emulated

LAN name is *man*, the second client is linked to that emulated LAN name by default. Router 4 is configured as shown in the following code:

```
interface atm 3/0
 atm pvc 1 0 5 qsaal
 atm pvc 2 0 16 ilmi
interface atm 3/0.1
 ip address 172.16.2.4 255.255.255.0
 lane server-bus tokenring mkt
 lane client tokenring mkt
interface atm 3/0.2
 ip address 172.16.0.4 255.255.255.0
 lane client tokenring
```

Multiple Token Ring ELANs with Restricted Membership Example

The following example, illustrated in Figure 8–2, configures a Cisco 7500 series router for three emulated LANS for Engineering, Manufacturing, and Marketing. The same components are assigned to the four routers as in the previous example. The ATM address prefixes and MAC addresses are also the same as in the previous example. However, this example restricts membership for the Engineering and Marketing emulated LANs. The LANE configuration server's database has explicit entries binding the ATM addresses of LANE clients to specified, named emulated LANs. In such cases, the client requests information from the configuration server about which emulated LAN it should join; the configuration server checks its database and replies to the client. Because the Manufacturing emulated LAN is unrestricted, any client not in the LANE configuration server's database is allowed to join it.

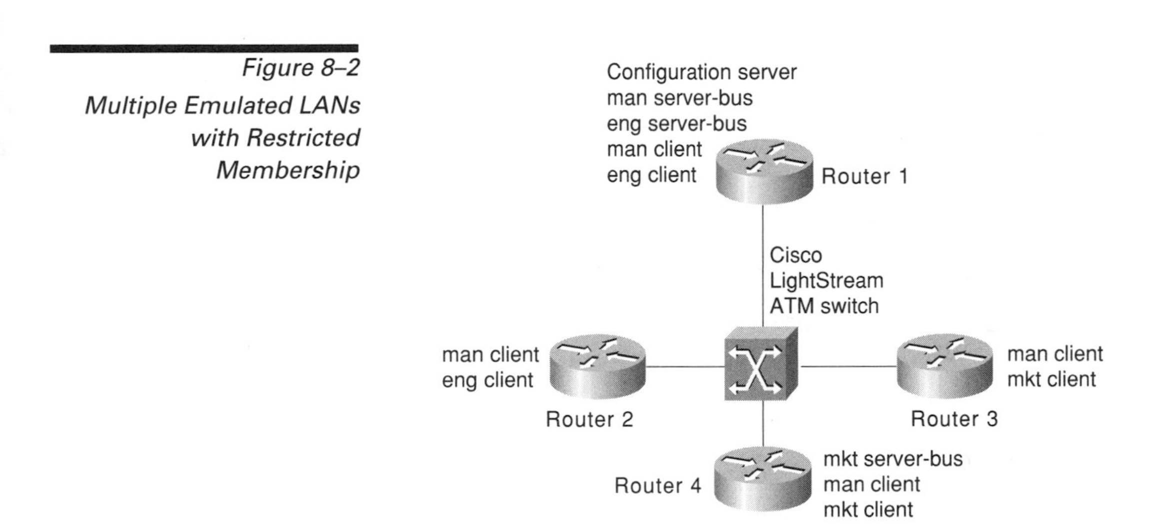

Figure 8–2
Multiple Emulated LANs with Restricted Membership

Router 1

Router 1 has the configuration server and its database, the server and broadcast-and-unknown server for the Manufacturing emulated LAN, the server and broadcast-and-unknown server for the Engineering emulated LAN, a client for Manufacturing, and a client for Engineering. It also has explicit database entries binding the ATM addresses of LANE clients to specified, named emulated LANs. Router 1 is configured as shown in the following code:

```
! The following lines name and configure the configuration server's database.
lane database example3
 name eng server-atm-address 39.000001415555121101020304.0800.200c.1001.02 restricted
 name eng local-seg-id 1000
 name man server-atm-address 39.000001415555121101020304.0800.200c.1001.01
 name man local-seg-id 2000
 name mkt server-atm-address 39.000001415555121101020304.0800.200c.4001.01 restricted
 name mkt local-seg-id 3000
 !
 ! The following lines add database entries binding specified client ATM
 ! addresses to emulated LANs. In each case, the Selector byte corresponds
 ! to the subinterface number on the specified router.
 ! The next command binds the client on Router 1's subinterface 2 to the eng ELAN.
 client-atm-address 39.0000014155551211.0800.200c.1000.02 name eng
 ! The next command binds the client on Router 2's subinterface 2 to the eng ELAN.
 client-atm-address 39.0000014155551211.0800.200c.2000.02 name eng
 ! The next command binds the client on Router 3's subinterface 2 to the mkt ELAN.
 client-atm-address 39.0000014155551211.0800.200c.3000.02 name mkt
 ! The next command binds the client on Router 4's subinterface 1 to the mkt ELAN.
 client-atm-address 39.0000014155551211.0800.200c.4000.01 name mkt
```

```
 default-name man
!
! The following lines bring up the configuration server and associate
! it with a database name.
interface atm 1/0
 atm pvc 1 0 5 qsaal
 atm pvc 2 0 16 ilmi
 lane config auto-config-atm-address
 lane config database example3
!
! The following lines configure the "man" server/broadcast-and-unknown server,
! and the client on atm subinterface 1/0.1. The client is assigned to the default
! emulated lan.
interface atm 1/0.1
 ip address 172.16.0.1 255.255.255.0
 lane server-bus tokenring man
 lane client tokenring
!
! The following lines configure the "eng" server/broadcast-and-unknown server
! and the client on atm subinterface 1/0.2. The configuration server assigns the
! client to the engineering emulated lan.
interface atm 1/0.2
 ip address 172.16.1.1 255.255.255.0
 lane server-bus tokenring eng
 lane client tokenring eng
```

Router 2

Router 2 is configured for a client of the Manufacturing emulated LAN and a client of the Engineering emulated LAN. Because the default emulated LAN name is *man*, the first client is linked to that emulated LAN name by default. Router 2 is configured as shown in the following example:

```
interface atm 1/0
 atm pvc 1 0 5 qsaal
 atm pvc 2 0 16 ilmi
! This client is not in the configuration server's database, so it will be
! linked to the "man" ELAN by default.
interface atm 1/0.1
 ip address 172.16.0.2 255.255.255.0
 lane client tokenring
! A client for the following interface is entered in the configuration
! server's database as linked to the "eng" ELAN.
interface atm 1/0.2
 ip address 172.16.1.2 255.255.255.0
 lane client tokenring eng
```

Router 3

Router 3 is configured for a client of the Manufacturing emulated LAN and a client of the Marketing emulated LAN. Because the default emulated LAN name is *man*, the first client is linked to that emulated LAN name by default. The second client is listed in the database as linked to the *mkt* emulated LAN. Router 3 is configured as shown in the following example:

```
interface atm 2/0
 atm pvc 1 0 5 qsaal
 atm pvc 2 0 16 ilmi
! The first client is not entered in the database, so it is linked to the
! "man" ELAN by default.
interface atm 2/0.1
 ip address 172.16.0.3 255.255.255.0
 lane client tokenring man
! The second client is explicitly entered in the configuration server's
! database as linked to the "mkt" ELAN.
interface atm 2/0.2
 ip address 172.16.2.3 255.255.255.0
 lane client tokenring mkt
```

Router 4

Router 4 has the server and broadcast-and-unknown server for the Marketing emulated LAN, a client for Marketing, and a client for Manufacturing. The first client is listed in the database as linked to the *mkt* emulated LANs. The second client is not listed in the database, but is linked to the *man* emulated LAN name by default. Router 4 is configured as shown in the following code:

```
interface atm 3/0
 atm pvc 1 0 5 qsaal
 atm pvc 2 0 16 ilmi
! The first client is explicitly entered in the configuration server's
! database as linked to the "mkt" ELAN.
interface atm 3/0.1
 ip address 172.16.2.4 255.255.255.0
 lane server-bus tokenring mkt
 lane client tokenring mkt
! The following client is not entered in the database, so it is linked to the
! "man" ELAN by default.
interface atm 3/0.2
 ip address 172.16.0.4 255.255.255.0
 lane client tokenring
```

TR-LANE with 2-Port Source-Route Bridging Example

The following example configures two Cisco 7500 series routers for one emulated Token-Ring LAN using source-route bridging, as illustrated in Figure 8–3. This example does not restrict membership in the emulated LANs.

Figure 8–3
2-Port Source-Route Bridging TR-LANE

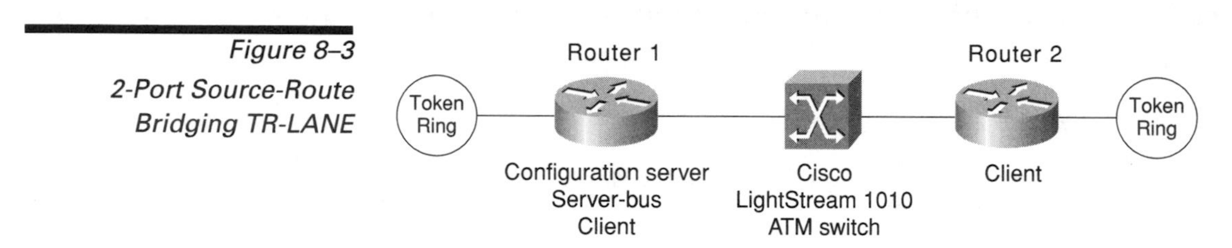

Router 1

Router 1 contains the configuration server, the server and broadcast-and-unknown server, and a client. Router 1 is configured as shown in the following example:

```
hostname Router1
!
! The following lines configure the database cisco_eng.
lane database cisco_eng
 name elan1 server-atm-address 39.020304050607080910111213.00000CA05B41.01
 name elan1 local-seg-id 2048
 default-name elan1
!
interface Ethernet0/0
 ip address 10.6.10.4 255.255.255.0
!
! The following lines configure a configuration server using the cisco_eng database on
! the interface. No IP address is needed since we are using source-route bridging.
interface ATM2/0
 no ip address
 atm pvc 1 0 5 qsaal
 atm pvc 2 0 16 ilmi
 lane config auto-config-atm-address
 lane config database cisco_eng
!
! The following lines configure the server-bus and the client on the subinterface and
! specify source-route bridging information.
interface ATM2/0.1 multipoint
 lane server-bus tokenring elan1
 lane client tokenring elan1
 source-bridge 2048 1 1
```

```
 source-bridge spanning
!
! The following lines configure source-route bridging on the Token Ring interface.
interface TokenRing3/0/0
 no ip address
 ring-speed 16
 source-bridge 1 1 2048
 source-bridge spanning
!
router igrp 65529
 network 10.0.0.0
```

Router 2

Router 2 contains only a client for the emulated LAN. Router 2 is configured as shown in the following code:

```
hostname Router2
!
interface Ethernet0/0
 ip address 10.6.10.5 255.255.255.0
!
! The following lines configure source-route bridging on the Token Ring interface.
interface TokenRing1/0
 no ip address
 ring-speed 16
 source-bridge 2 2 2048
 source-bridge spanning
!
! The following lines set up the signaling and ILMI PVCs.
interface ATM2/0
 no ip address
 atm pvc 1 0 5 qsaal
 atm pvc 2 0 16 ilmi
!
! The following lines set up a client on the subinterface and configure
! source-route bridging.
interface ATM2/0.1 multipoint
 ip address 1.1.1.2 255.0.0.0
 lane client tokenring elan1
 source-bridge 2048 2 2
 source-bridge spanning
!
router igrp 65529
 network 10.0.0.0
```

TR-LANE with Multiport Source-Route Bridging Example

The following example configures two Cisco 7500 series routers for one emulated Token-Ring LAN using source-route bridging, as illustrated in Figure 8–4. Because each router connects to three rings (the two Token Rings and the emulated LAN "ring"), a virtual ring must be configured on the router. This example does not restrict membership in the emulated LANs.

Figure 8–4 Multiport Source-Route Bridged Token Ring Emulated LAN

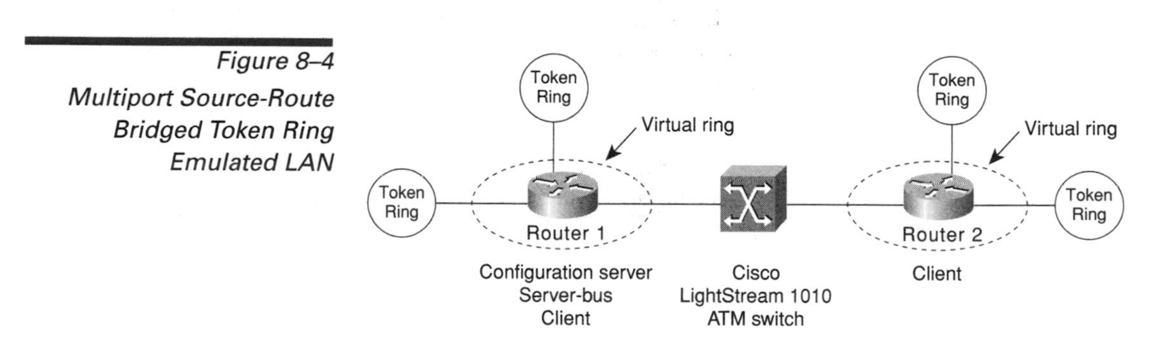

Router 1

Router 1 contains the configuration server, the server and broadcast-and-unknown server, and a client. Router 1 is configured as shown in the following example:

```
hostname Router1
!
! The following lines configure the database with the information about the
! elan1 emulated Token Ring LAN.
lane database cisco_eng
 name elan1 server-atm-address 39.020304050607080910111213.00000CA05B41.01
 name elan1 local-seg-id 2048
 default-name elan1
!
! The following line configures virtual ring 256 on the router.
source-bridge ring-group 256
!
interface Ethernet0/0
 ip address 10.6.10.4 255.255.255.0
!
! The following lines configure the configuration server to use the cisco_eng database.
! The Signalling and ILMI PVCs are also configured.
interface ATM2/0
 no ip address
 atm pvc 1 0 5 qsaal
```

```
 atm pvc 2 0 16 ilmi
 lane config auto-config-atm-address
 lane config database cisco_eng
!
! The following lines configure the server and broadcast-and-unknown server and a client
! on the interface. The lines also specify source-route bridging information.
interface ATM2/0.1 multipoint
 lane server-bus tokenring elan1
 lane client tokenring elan1
 source-bridge 2048 5 256
 source-bridge spanning
!
! The following lines configure the Token Ring interfaces.
interface TokenRing3/0
 no ip address
 ring-speed 16
 source-bridge 1 1 256
 source-bridge spanning
interface TokenRing3/1
 no ip address
 ring-speed 16
 source-bridge 2 2 256
 source-bridge spanning
!
router igrp 65529
 network 10.0.0.0
```

Router 2

Router 2 contains only a client for the emulated LAN. Router 2 is configured as shown in the following code:

```
hostname Router2
!
! The following line configures virtual ring 512 on the router.
source-bridge ring-group 512
!
interface Ethernet0/0
 ip address 10.6.10.5 255.255.255.0
!
! The following lines configure the Token Ring interfaces.
interface TokenRing1/0
 no ip address
 ring-speed 16
 source-bridge 3 3 512
 source-bridge spanning
interface TokenRing1/1
 no ip address
 ring-speed 16
```

```
 source-bridge 4 4 512
 source-bridge spanning
!
! The following lines configure the signaling and ILMI PVCs.
interface ATM2/0
 no ip address
 atm pvc 1 0 5 qsaal
 atm pvc 2 0 16 ilmi
!
! The following lines configure the client. Source-route bridging is also configured.
interface ATM2/0.1 multipoint
 ip address 1.1.1.2 255.0.0.0
 lane client tokenring elan1
 source-bridge 2048 6 512
 source-bridge spanning
!
router igrp 65529
 network 10.0.0.0
```

Routing Between Token Ring and Ethernet Emulated LANs Example

This example, shown in Figure 8–5, configures routing between a Token Ring emulated LAN (*trelan*) and an Ethernet emulated LAN (*ethelan*) on the same ATM interface. Router 1 contains the LANE configuration server, a LANE server and broadcast-and-unknown server for each emulated LAN, and a client for each emulated LAN. Router 2 contains a client for *trelan* (Token Ring); Router 3 contains a client for *ethelan* (Ethernet).

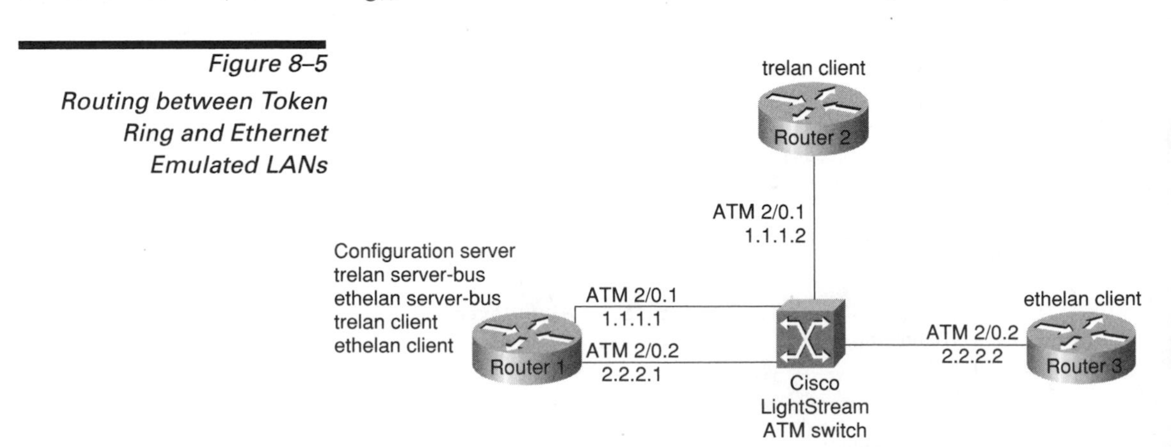

Figure 8–5
Routing between Token Ring and Ethernet Emulated LANs

Router 1

Router 1 contains the LANE configuration server, a LANE server and broadcast-and-unknown server for each emulated LAN, and a client for each emulated LAN. Router 1 is configured as shown in the following example:

```
hostname router1
!
! The following lines name and configures the configuration server's database.
! The server addresses for trelan and ethelan and the ELAN ring number for
! trelan are entered into the database. The default ELAN is trelan.
lane database cisco_eng
 name trelan server-atm-address 39.020304050607080910111213.00000CA05B41.01
 name trelan local-seg-id 2048
 name ethelan server-atm-address 39.020304050607080910111213.00000CA05B41.02
 default-name trelan
!
! The following lines enable the configuration server and associate it
! with the cisco_eng database.
interface ATM2/0
 no ip address
 atm pvc 1 0 5 qsaal
 atm pvc 2 0 16 ilmi
 lane config auto-config-atm-address
 lane config database cisco_eng
!
! The following lines configure the tokenring LES/BUS and LEC for trelan
! on subinterface atm2/0.1 and assign an IP address to the subinterface.
interface ATM2/0.1 multipoint
 ip address 10.1.1.1 255.255.255.0
 lane server-bus tokenring trelan
 lane client tokenring trelan
!
! The following lines configure the Ethernet LES/BUS and LEC for ethelan
! on subinterface atm2/0.2 and assign an IP address to the subinterface.
interface ATM2/0.2 multipoint
 ip address 20.2.2.1 255.255.255.0
 lane server-bus ethernet ethelan
 lane client ethernet ethelan
!
! The following lines configure the IGRP routing protocol to enable routing
! between ELANS.
router igrp 1
 network 10.0.0.0
 network 20.0.0.0
```

Router 2

Router 2 contains a client for *trelan* (Token Ring). Router 2 is configured as follows:

```
hostname router2
!
! The following lines set up the signaling and ILMI PVCs for the interface.
interface ATM2/0
 no ip address
 no keepalive
 atm pvc 1 0 5 qsaal
 atm pvc 2 0 16 ilmi
!
! The following lines configure a Token Ring LEC on atm2/0.1 and assign
! an IP address to the subinterface.
interface ATM2/0.1 multipoint
 ip address 10.1.1.2 255.255.255.0
 lane client tokenring trelan
!
! The following lines configure the IGRP routing protocol to enable routing
! between ELANS.
router igrp 1
 network 10.0.0.0
 network 20.0.0.0
```

Router 3

Router 3 contains a client for *ethelan* (Ethernet). Router 3 is configured as follows:

```
hostname router3
!
! The following lines set up the signaling and ILMI PVCs for the interface.
interface ATM2/0
 no ip address
 no ip mroute-cache
 atm pvc 1 0 5 qsaal
 atm pvc 2 0 16 ilmi
!
! The following lines configure an Ethernet LEC on atm2/0.1 and assign
! an IP address to the subinterface.
interface ATM2/0.1 multipoint
 ip address 20.2.2.2 255.255.255.0
 lane client ethernet ethelan
!
! The following lines configure the IGRP routing protocol to enable routing
! between ELANS.
router igrp 1
 network 10.0.0.0
 network 20.0.0.0
```

CHAPTER 9

LAN Emulation Commands

This chapter describes the commands available to configure LAN emulation (LANE) in Cisco 7500 series routers, Cisco 4700 routers, and Cisco 4500 routers that contain an ATM Interface Processor (AIP) or an NP-1A ATM Network Processor Module (NPM) and are connected to a Cisco ATM switch.

NOTES

In Cisco IOS Release 11.3, all commands supported on the Cisco 7500 series routers are also supported on Cisco 7000 series routers. Because some LANE commands are used often and others are used very rarely, the command descriptions identify the commands you are most likely to use. See the "Usage Guidelines" section for the following indicator: "This command is ordinarily used."

CLEAR ATM VC

To release a specified switched virtual circuit (SVC), use the **clear atm vc** EXEC command.

clear atm vc *vcd*

Syntax	*Description*
vcd	Virtual channel descriptor of the channel to be released.

Command Mode

EXEC

Usage Guidelines

This command first appeared in Cisco IOS Release 11.0.

For multicast or control VCCs, this command causes the LANE client to exit and rejoin an emulated LAN.

For data VCCs, this command also removes the associated LANE Address Resolution Protocol (LE ARP) table entries.

Example

The following example releases SVC 1024:

```
clear atm vc 1024
```

CLEAR LANE LE-ARP

To clear the dynamic LANE Address Resolution Protocol (LE ARP) table or a single LE ARP entry of the LANE client configured on the specified subinterface or emulated LAN, use the **clear lane le-arp** EXEC command.

clear lane le-arp [**interface** *slot/port*[*.subinterface-number*] | **name** *elan-name*] [**mac-address** *mac-address* | **route-desc segment** *segment-number* **bridge** *bridge-number*] (for the Cisco 7500 series)

clear lane le-arp [**interface** *number*[*.subinterface-number*] | **name** *elan-name*] [**mac-address** *mac-address* | **route-desc segment** *segment-number* **bridge** *bridge-number*]] (for the Cisco 4500 and 4700 routers)

Syntax	*Description*
interface *slot/port* [*.subinterface-number*]	(Optional) Interface or subinterface for the LANE client whose LE ARP table or entry is to be cleared for the Cisco 7500 series routers. The space between the **interface** keyword and the *slot* argument is optional.
interface *number* [*.subinterface-number*]	(Optional) Interface or subinterface for the LANE client whose LE ARP table or entry is to be cleared for the Cisco 4500 or 4700 routers. The space between the **interface** keyword and the *number* argument is optional.
name *elan-name*	(Optional) Name of the emulated LAN for the LANE client whose LE ARP table or entry is to be cleared. Maximum length is 32 characters.

Syntax	*Description*
mac-address *mac-address*	(Optional) Media access control (MAC) address of the entry to be cleared from the LE ARP table.
route-desc segment *segment-number*	(Optional) LANE segment number. The segment number ranges from 1 to 4095.
bridge-number	(Optional) Bridge number that is contained in the route descriptor. The bridge number ranges from 1 to 15.

Command Mode

EXEC

Usage Guidelines

This command first appeared in Cisco IOS Release 11.0.

This command removes dynamic LE ARP table entries only. It does not remove static LE ARP table entries.

If you do not specify an interface or an emulated LAN, this command clears all the LE ARP tables of any LANE client in the router.

If you specify a major interface (not a subinterface), this command clears all the LE ARP tables of every LANE client on all the subinterfaces of that interface.

This command also removes the fast-cache entries built from the LE ARP entries.

Examples

The following example clears all the LE ARP tables for all clients on the router:

```
clear lane le-arp
```

The following example clears all the LE ARP tables for all LANE clients on all the subinterfaces of interface 1/0:

```
clear lane le-arp interface 1/0
```

The following example clears the entry corresponding to MAC address 0800.AA00.0101 from the LE ARP table for the LANE client on the emulated LAN *red*:

```
clear lane le-arp name red 0800.aa00.0101
```

The following example clears all dynamic entries from the LE ARP table for the LANE client on the emulated LAN *red*:

```
clear lane le-arp name red
```

The following example clears the dynamic entry from the LE ARP table for the LANE client on segment number 1, bridge number 1 in the emulated LAN *red*:

```
clear lane le-arp name red route-desc segment 1 bridge 1
```

NOTES

MAC addresses are written in the same dotted notation for the **clear lane le-arp** command as they are for the global IP **arp** command.

CLEAR LANE SERVER

To force a LANE server to drop a client and allow the LANE configuration server to assign the client to another emulated LAN, use the **clear lane server** EXEC command.

clear lane server {**interface** *slot/port* [*.subinterface-number*] | **name** *elan-name*} [**mac-address** *mac-address* | **client-atm-address** *atm-address* | **lecid** *lane-client-id* | **route-desc segment** *segment-number* **bridge** *bridge-number*]
(for the Cisco 7500 series)

clear lane server {**interface** *number* [*.subinterface-number*] | **name** *elan-name*} [**mac-address** *mac-address* | **client-atm-address** *atm-address* | **lecid** *lecid* | **route-desc segment** *segment-number* **bridge** *bridge-number*]
(for the Cisco 4500 and 4700 routers)

Syntax	*Description*
interface *slot/port* [*.subinterface-number*]	Interface or subinterface where the LANE server is configured for the Cisco 7500 series. The space between the **atm** keyword and the *slot* argument is optional.
interface *number* [*.subinterface-number*]	Interface or subinterface where the LANE server is configured for the Cisco 4500 or 4700 routers. The space between the **atm** keyword and the *number* argument is optional.
name *elan-name*	Name of the emulated LAN on which the LANE server is configured. Maximum length is 32 characters.
mac-address *mac-address*	(Optional) Keyword and LANE client's MAC address.
client-atm-address *atm-address*	(Optional) Keyword and LANE client's ATM address.

Syntax	*Description*
lecid *lane-client-id*	(Optional) Keyword and LANE client ID. The LANE client ID is a value between 1 and 4096.
route-desc segment *segment-number*	(Optional) Keywords and LANE segment number. The segment number ranges from 1 to 4095.
bridge *bridge-number*	(Optional) Keyword and bridge number that is contained in the route descriptor. The bridge number ranges from 1 to 15.

Command Mode

EXEC

Usage Guidelines

This command first appeared in Cisco IOS Release 11.0.

After changing the bindings on the configuration server, enter this command on the LANE server to force the client to leave one emulated LAN. The LANE server will drop the Control Direct and Control Distribute VCCs to the LANE client. The client will then ask the LANE configuration server for the location of the LANE server of the emulated LAN it should join. If no LANE client is specified, all LANE clients attached to the LANE server are dropped.

Example

The following example forces all the LANE clients on the emulated LAN *red* to be dropped. The next time they try to join, they will be forced to join a different emulated LAN.

```
clear lane server name red
```

Related Commands

You can use the master index or search online to find documentation of related commands.

client-atm-address name
lane database
mac-address name
show lane server

CLIENT-ATM-ADDRESS NAME

To add a LANE client address entry to the configuration server's configuration database, use the **client-atm-address** database configuration command. To remove a client address entry from the table, use the **no** form of this command.

client-atm-address *atm-address-template* **name** *elan-name*
no client-atm-address *atm-address-template*

Syntax	*Description*
atm-address-template	Template that explicitly specifies an ATM address or a specific part of an ATM address and uses wildcard characters for other parts of the ATM address, making it easy and convenient to specify multiple addresses matching the explicitly specified part. Wildcard characters can replace any nibble or group of nibbles in the prefix, the end-system identifier (ESI), or the selector fields of the ATM address.
elan-name	Name of the emulated LAN. Maximum length is 32 characters.

Defaults

No address and no emulated LAN name are provided.

Command Mode

Database configuration

Usage Guidelines

This command first appeared in Cisco IOS Release 11.0.

This command is ordinarily used.

The effect of this command is to bind any client whose address matches the specified template into the specified emulated LAN. When a client comes up, it consults the LANE configuration server, which responds with the ATM address of the LANE server for the emulated LAN. The client then initiates join procedures with the LANE server.

Before this command is used, the emulated LAN specified by the *elan-name* argument must have been created in the configuration server's database by use of the **name server-atm-address** command.

If an existing entry in the configuration server's database binds the LANE client ATM address to a different emulated LAN, the new command is rejected.

This command affects only the bindings in the named configuration server database. It has no effect on the LANE components themselves.

See the **lane database** command for information about creating the database, and the **name server-atm-address** command for information about binding the emulated LAN's name to the server's ATM address.

The **client-atm-address name** command is a subcommand of the global **lane database** command.

ATM Addresses. A LANE ATM address has the same syntax as a network service access point (NSAP) but it is not a network-level address. It consists of the following:

- A 13-byte prefix that includes the following fields defined by the ATM Forum:
    - AFI (Authority and Format Identifier) field (1 byte), DCC (Data Country Code) or ICD (International Code Designator) field (2 bytes), DFI field (Domain Specific Part Format Identifier) (1 byte), Administrative Authority field (3 bytes), Reserved field (2 bytes), Routing Domain field (2 bytes), and the Area field (2 bytes)
- A 6-byte end-system identifier (ESI)
- A 1-byte selector field

Address Templates. LANE ATM address templates can use two types of wildcards: an asterisk (*) to match any single character (nibble), and an ellipsis (...) to match any number of leading, middle, or trailing characters. The values of the characters replaced by wildcards come from the automatically assigned ATM address.

In LANE, a *prefix template* explicitly matches the prefix but uses wildcards for the ESI and selector fields. An *ESI template* explicitly matches the ESI field but uses wildcards for the prefix and selector.

In our implementation of LANE, the prefix corresponds to the switch, the ESI corresponds to the ATM interface, and the selector field corresponds to the specific subinterface of the interface.

Examples

The following example uses an ESI template to specify the part of the ATM address corresponding to the interface. This example allows any client on any subinterface of the interface that corresponds to the displayed ESI value, no matter which switch the router is connected to, to join the *engineering* emulated LAN:

```
client-atm-address ...0800.200C.1001.** name engineering
```

The following example uses a prefix template to specify the part of the ATM address corresponding to the switch. This example allows any client on a subinterface of any interface connected to the switch that corresponds to the displayed prefix to join the *marketing* emulated LAN:

```
client-atm-address 47.000014155551212f.00.00... name marketing
```

Related Commands

You can use the master index or search online to find documentation of related commands.

default-name
lane database
mac-address name
name server-atm-address

DEFAULT-NAME

To provide an emulated LAN name in the configuration server's database for those client MAC addresses and client ATM addresses that do not have explicit emulated LAN name bindings, use the **default-name** database configuration command. To remove the default name, use the **no** form of this command.

default-name *elan-name*
no default-name

Syntax	*Description*
elan-name	Default emulated LAN name for any LANE client MAC address or LANE client ATM address not explicitly bound to any emulated LAN name. Maximum length is 32 characters.

Default

No name is provided.

Command Mode

Database configuration

Usage Guidelines

This command first appeared in Cisco IOS Release 11.0.

This command is ordinarily used.

This command affects only the bindings in the configuration server's database. It has no effect on the LANE components themselves.

The named emulated LAN must already exist in the configuration server's database before this command is used. If the default name-to-emulated LAN name binding already exists, the new binding replaces it.

The **default-name** command is a subcommand of the global **lane database** command.

Example

The following example specifies the emulated Token Ring LAN *man* as the default emulated LAN. Because none of the emulated LANs are restricted, clients are assigned to whichever emulated LAN they request. Clients that do not request a particular emulated LAN will be assigned to the *man* emulated LAN.

```
lane database example2
 name eng server-atm-address 39.000001415555121101020304.0800.200c.1001.02
 name eng local-seg-id 1000
 name man server-atm-address 39.000001415555121101020304.0800.200c.1001.01
 name man local-seg-id 2000
 name mkt server-atm-address 39.000001415555121101020304.0800.200c.4001.01
 name mkt local-seg-id 3000
 default-name man
```

Related Commands

You can use the master index or search online to find documentation of related commands.

client-atm-address name
lane database
mac-address name
name server-atm-address

LANE AUTO-CONFIG-ATM-ADDRESS

To specify that the configuration server ATM address is computed by our automatic method, use the **lane auto-config-atm-address** interface configuration command. To remove the previously assigned ATM address, use the **no** form of this command.

lane [config] auto-config-atm-address
no lane [config] auto-config-atm-address

Syntax	*Description*
config	(Optional) When the **config** keyword is used, this command applies only to the LANE configuration server (LECS). You are telling the LECS to use the auto-computed LECS address.

Default

No specific ATM address is set.

Command Mode

Interface configuration

Usage Guidelines

This command first appeared in Cisco IOS Release 11.0.

When the **config** keyword is not present, this command causes the LANE server and LANE client on the subinterface to use the automatically assigned ATM address for the configuration server.

When the **config** keyword is present, this command assigns the automatically generated ATM address to the configuration server (LECS) configured on the interface. Multiple commands that assign ATM addresses to the LANE configuration server can be issued on the same interface to assign different ATM addresses to the configuration server. Commands that assign ATM addresses to the LANE configuration server include **lane auto-config-atm-address, lane config-atm-address,** and **lane fixed-config-atm-address.**

Examples

The following example associates the LANE configuration server with the database named *network1* and specifies that the configuration server's ATM address will be assigned by our automatic method:

```
lane database network1
 name eng server-atm-address 39.020304050607080910111213.0800.AA00.1001.02
 name mkt server-atm-address 39.020304050607080910111213.0800.AA00.4001.01
interface atm 1/0
 lane config database network1
 lane config auto-config-atm-address
```

The following example causes the LANE server and LANE client on the subinterface to use the automatically assigned ATM address to communicate with the configuration server:

```
interface atm 2/0.1
 ip address 172.16.0.4 255.255.255.0
 lane client ethernet
 lane server-bus ethernet eng
 lane auto-config-atm-address
```

Related Commands

You can use the master index or search online to find documentation of related commands.

lane config-atm-address
lane database
lane fixed-config-atm-address

LANE BUS-ATM-ADDRESS

To specify an ATM address, and thus override the automatic ATM address assignment, for the broadcast-and-unknown server on the specified subinterface, use the **lane**

bus-atm-address interface configuration command. To remove the ATM address previously specified for the broadcast-and-unknown server on the specified subinterface and thus revert to the automatic address assignment, use the **no** form of this command.

lane bus-atm-address *atm-address-template*
no lane bus-atm-address [*atm-address-template*]

Syntax	*Description*
atm-address-template	ATM address or a template in which wildcard characters are replaced by any nibble or group of nibbles of the prefix bytes, the end-system identifier (ESI) bytes, or the selector byte of the automatically assigned ATM address.

Default

For the broadcast-and-unknown server, the default is automatic ATM address assignment.

Command Mode

Interface configuration

Usage Guidelines

This command first appeared in Cisco IOS Release 11.0.

When applied to a broadcast-and-unknown server, this command overrides automatic ATM address assignment for the broadcast-and-unknown server. When applied to a LANE client, this command gives the client the ATM address of the broadcast-and-unknown server. The client will use this address rather than sending LE ARP requests for the broadcast address.

When applied to a selected interface, but with a different ATM address than was used previously, this command replaces the broadcast-and-unknown server's ATM address.

ATM Addresses. A LANE ATM address has the same syntax as an NSAP (but it is not a network-level address). It consists of the following:

- A 13-byte prefix that includes the following fields defined by the ATM Forum:
    - AFI (Authority and Format Identifier) field (1 byte)
    - DCC (Data Country Code) or ICD (International Code Designator) field (2 bytes)

  - DFI field (Domain Specific Part Format Identifier) (1 byte)
  - Administrative Authority field (3 bytes)
  - Reserved field (2 bytes)
  - Routing Domain field (2 bytes)
  - Area field (2 bytes)
- A 6-byte end-system identifier (ESI)
- A 1-byte selector field

Address Templates. LANE ATM address templates can use two types of wildcards: an asterisk (*) to match any single character (nibble), and an ellipsis (...) to match any number of leading, middle, or trailing characters. The values of the characters replaced by wildcards come from the automatically assigned ATM address.

The values of the digits that are replaced by wildcards come from the automatic ATM assignment method.

In LANE, a *prefix template* explicitly matches the prefix but uses wildcards for the ESI and selector fields. An *ESI template* explicitly matches the ESI field but uses wildcards for the prefix and selector.

In our implementation of LANE, the prefix corresponds to the switch, the ESI corresponds to the ATM interface, and the Selector field corresponds to the specific subinterface of the interface.

Examples

The following example uses an ESI template to specify the part of the ATM address corresponding to the interface; the remaining values in the ATM address come from automatic assignment:

```
lane bus-atm-address ...0800.200C.1001.**
```

The following example uses a prefix template to specify the part of the ATM address corresponding to the switch; the remaining values in the ATM address come from automatic assignment:

```
lane bus-atm-address 45.000014155551212f.00.00...
```

Related Commands

You can use the master index or search online to find documentation of related commands.

lane server-bus

LANE CLIENT

To activate a LANE client on the specified subinterface, use the **lane client** interface configuration command. To remove a previously activated LANE client on the subinterface, use the **no** form of this command.

lane client {**ethernet** | **tokenring**} [*elan-name*]
no lane client [{**ethernet** | **tokenring**} [*elan-name*]]

Syntax	*Description*
ethernet	Identifies the emulated LAN attached to this subinterface as an Ethernet ELAN.
tokenring	Identifies the emulated LAN attached to this subinterface as a Token Ring ELAN.
elan-name	(Optional) Name of the emulated LAN. This argument is optional because the client obtains its emulated LAN name from the configuration server. The maximum length of the name is 32 characters.

Default

No LANE clients are enabled on the interface.

Command Mode

Interface configuration

Usage Guidelines

This command first appeared in Cisco IOS Release 11.0.

This command is ordinarily used.

If a **lane client** command has already been entered on the subinterface for a different emulated LAN, then the client initiates termination procedures for that emulated LAN and joins the new emulated LAN.

If you do not provide an *elan-name* value, the client contacts the server to find which emulated LAN to join. If you do provide an emulated LAN name, the client consults the configuration server to ensure that no conflicting bindings exist.

Example

The following example enables a Token Ring LANE client on an interface:

```
lane client tokenring
```

Related Commands

You can use the master index or search online to find documentation of related commands.

lane client-atm-address

LANE CLIENT-ATM-ADDRESS

To specify an ATM address, and thus override the automatic ATM address assignment, for the LANE client on the specified subinterface, use the **lane client-atm-address** interface configuration command. To remove the ATM address previously specified for the LANE client on the specified subinterface and thus revert to the automatic address assignment, use the **no** form of this command.

lane client-atm-address *atm-address-template*
no client-atm-address [*atm-address-template*]

Syntax	*Description*
atm-address-template	ATM address or a template in which wildcard characters are replaced by any nibble or group of nibbles of the prefix bytes, the ESI bytes, or the selector byte of the automatically assigned ATM address.

Default

Automatic ATM address assignment.

Command Mode

Interface configuration

Usage Guidelines

This command first appeared in Cisco IOS Release 11.0.

Use of this command on a selected subinterface, but with a different ATM address than was used previously, replaces the LANE client's ATM address.

ATM Addresses. A LANE ATM address has the same syntax as an NSAP (but it is not a network-level address). It consists of the following:

- A 13-byte prefix that includes the following fields defined by the ATM Forum:
    - AFI (Authority and Format Identifier) field (1 byte)
    - DCC (Data Country Code) or ICD (International Code Designator) field (2 bytes)
    - DFI field (Domain Specific Part Format Identifier) (1 byte)
    - Administrative Authority field (3 bytes)
    - Reserved field (2 bytes)
    - Routing Domain field (2 bytes)
    - Area field (2 bytes)
- A 6-byte end-system identifier (ESI)
- A 1-byte selector field

Address Templates. LANE ATM address templates can use two types of wildcards: an asterisk (*) to match any single character (nibble), and an ellipsis (...) to match any number of leading, middle, or trailing characters. The values of the characters replaced by wildcards come from the automatically assigned ATM address.

In LANE, a *prefix template* explicitly matches the ATM address prefix but uses wildcards for the ESI and selector fields. An *ESI template* explicitly matches the ESI field but uses wildcards for the prefix and selector.

In our implementation of LANE, the prefix corresponds to the switch, the ESI corresponds to the ATM interface, and the selector field corresponds to the specific subinterface of the interface.

For a discussion of Cisco's method of automatically assigning ATM addresses, refer to the "Configuring LAN Emulation" in Chapter 8.

Examples

The following example uses an ESI template to specify the part of the ATM address corresponding to the interface; the remaining parts of the ATM address come from automatic assignment:

```
lane client-atm-address...0800.200C.1001.**
```

The following example uses a prefix template to specify the part of the ATM address corresponding to the switch; the remaining parts of the ATM address come from automatic assignment:

```
lane client-atm-address 47.000014155551212f.00.00...
```

Related Commands

You can use the master index or search online to find documentation of related commands.

lane client

LANE CONFIG-ATM-ADDRESS

To specify a configuration server's ATM address explicitly, use the **lane config-atm-address** interface configuration command. To remove an assigned ATM address, use the **no** form of this command.

lane [config] config-atm-address *atm-address-template*
no lane [config] config-atm-address *atm-address-template*

Syntax	*Description*
config	(Optional) When the **config** keyword is used, this command applies only to the LANE configuration server (LECS). You are telling the LECS to use the 20-byte address that you explicitly entered.
atm-address-template	ATM address or a template in which wildcard characters are replaced by any nibble or group of nibbles of the prefix bytes, the ESI bytes, or the selector byte of the automatically assigned ATM address.

Default

No specific ATM address or method is set.

Command Mode

Interface configuration

Usage Guidelines

This command first appeared in Cisco IOS Release 11.0.

If the **config** keyword is not present, this command causes the LANE server and LANE client on the subinterface to use the specified ATM address for the configuration server.

When the **config** keyword is present, this command adds an ATM address to the configuration server configured on the interface. A LANE configuration server can listen on multiple ATM addresses. Multiple commands that assign ATM addresses to the LANE configuration server can be issued on the same interface to assign different ATM addresses to the LANE configuration server.

ATM Addresses. A LANE ATM address has the same syntax as an NSAP (but it is not a network-level address). It consists of the following:

- A 13-byte prefix that includes the following fields defined by the ATM Forum:
    - AFI (Authority and Format Identifier) field (1 byte)
    - DCC (Data Country Code) or ICD (International Code Designator) field (2 bytes)
    - DFI field (Domain Specific Part Format Identifier) (1 byte)

- ○ Administrative Authority field (3 bytes)
- ○ Reserved field (2 bytes)
- ○ Routing Domain field (2 bytes)
- ○ Area field (2 bytes)

- A 6-byte end-system identifier (ESI)
- A 1-byte selector field

Address Templates. LANE ATM address templates can use two types of wildcards: an asterisk (*) to match any single character (nibble), and an ellipsis (...) to match any number of leading, middle, or trailing characters. The values of the characters replaced by wildcards come from the automatically assigned ATM address.

In LANE, a *prefix template* explicitly matches the ATM address prefix but uses wildcards for the ESI and selector fields. An *ESI template* explicitly matches the ESI field but uses wildcards for the prefix and selector.

In our implementation of LANE, the prefix corresponds to the switch prefix, the ESI corresponds to a function of ATM interface's MAC address, and the Selector field corresponds to the specific subinterface of the interface.

Examples

The following example associates the LANE configuration server with the database named *network1* and explicitly specifies the configuration server's ATM address:

```
lane database network1
 name eng server-atm-address 39.020304050607080910111213.0800.AA00.1001.02
 name mkt server-atm-address 39.020304050607080910111213.0800.AA00.4001.01
interface atm 1/0
 lane config database network1
 lane config config-atm-address 39.020304050607080910111213.0800.AA00.3000.00
```

The following example causes the LANE server and LANE client on the subinterface to use the explicitly specified ATM address to communicate with the configuration server:

```
interface atm 2/0.1
 ip address 172.16.0.4 255.255.255.0
 lane client ethernet
 lane server-bus ethernet eng
 lane config-atm-address 39.020304050607080910111213.0800.AA00.3000.00
```

Related Commands

You can use the master index or search online to find documentation of related commands.

lane auto-config-atm-address
lane config database
lane database
lane fixed-config-atm-address

LANE CONFIG DATABASE

To associate a named configuration table (database) with the configuration server on the selected ATM interface, use the **lane config database** interface configuration command. To remove the association between a named database and the configuration server on the specified interface, use the **no** form of this command.

lane config database *database-name*
no lane config

Syntax	*Description*
database-name	Name of the LANE database.

Default

No configuration server is defined, and no database name is provided.

Command Mode

Interface configuration

Usage Guidelines

This command first appeared in Cisco IOS Release 11.0.

This command is valid only on a major interface, not a subinterface, because only one LANE configuration server can exist per interface.

The named database must exist before the **lane config database** command is entered. Refer to the **lane database** command for more information.

Multiple **lane config database** commands cannot be entered multiple times on the same interface. You must delete an existing association by using the **no** form of this command before you can create a new association on the specified interface.

Activating a LANE configuration server requires the **lane config database** command and one of the following commands: **lane config fixed-config-atm-address, lane config auto-config-atm-address**, or **lane config config-atm-address.**

Example

The following example associates the LANE configuration server with the database named *network1* and specifies that the configuration server's ATM address will be assigned by our automatic method:

```
lane database network1
 name eng server-atm-address 39.020304050607080910111213.0800.AA00.1001.02
 name mkt server-atm-address 39.020304050607080910111213.0800.AA00.4001.01
interface atm 1/0
 lane config database network1
 lane config auto-config-atm-address
```

Related Commands

You can use the master index or search online to find documentation of related commands.

lane auto-config-atm-address
lane config-atm-address
lane database
lane fixed-config-atm-address

LANE DATABASE

To create a named configuration database that can be associated with a configuration server, use the **lane database** global configuration command. To delete the database, use the **no** form of this command.

lane database *database-name*
no lane database *database-name*

Syntax	*Description*
database-name	Database name (32 characters maximum).

Default

No name is provided.

Command Mode

Global configuration

Usage Guidelines

This command first appeared in Cisco IOS Release 11.0.

Use of the **lane database** command places you in database configuration mode, in which you can use the **client-atm-address name, default name, mac-address name, name restricted, name unrestricted, name new-name,** and **name server-atm-address** commands to create entries in the specified database. When you are finished creating entries, type **^Z** or **exit** to return to global configuration mode.

Example

The following example creates the database named *network1* and associates it with the configuration server on interface ATM 1/0:

```
lane database network1
 name eng server-atm-address 39.020304050607080910111213.0800.AA00.1001.02
 name mkt server-atm-address 39.020304050607080910111213.0800.AA00.4001.01
 default-name eng
interface atm 1/0
 lane config database network1
 lane config auto-config-atm-address
```

Related Commands

You can use the master index or search online to find documentation of related commands.

client-atm-address name
default-name
lane config database
mac-address name
name server-atm-address
name new-name

LANE FIXED-CONFIG-ATM-ADDRESS

To specify that the fixed configuration server ATM address assigned by the ATM Forum will be used, use the **lane fixed-config-atm-address** interface configuration command. To specify that the fixed ATM address is not used, use the **no** form of this command.

lane [**config**] **fixed-config-atm-address**
no lane [**config**] **fixed-config-atm-address**

Syntax	*Description*
config	(Optional) When the **config** keyword is used, this command applies only to the LANE configuration server (LECS). You are telling the LECS to use the well-known, ATM Forum, LEC address.

Default

No specific ATM address or method is set.

Command Mode

Interface configuration

Usage Guidelines

This command first appeared in Cisco IOS Release 11.0.

When the **config** keyword is not present, this command causes the LANE server and LANE client on the subinterface to use that ATM address, rather than the ATM address provided by the ILMI, to locate the configuration server.

When the **config** keyword is present, and the LECS is already up and running, please be aware of the following scenarios:

- If you configure the LECS with only the well-known address, the LECS will not participate in the SSRP, act as a "standalone" master, and only listen on the well-known LECS address. This scenario is ideal if you want a "standalone" LECS that does not participate in SSRP, and you would like to listen to only the well-known address.

- If only the well-known address is already assigned, and you assign at least one other address to the LECS, (additional addresses are assigned using the **lane config auto-config-atm-address** command or the **lane config config-atm-address command**) the LECS will participate in the SSRP and act as the master or slave based on the normal SSRP rules. This scenario is ideal if you would like the LECS to participate in SSRP, and you would like to make the master LECS listen on the well-known address.
- If the LECS is participating in SSRP, has more than one address (one of which is the well-known address), and all the addresses but the well-known address is removed, the LECS will declare itself the master and stop participating in SSRP completely.
- If the LECS is operating as an SSRP slave, and it has the well-known address configured, it will not listen on the well-known address unless it becomes the master.
- If you want the LECS to assume the well-known address only when it becomes the master, configure the LECS with the well-known address and at least one other address.

When you use this command with the **config** keyword, and the LAN Emulation Configuration Server (LECS) is a master, the master will listen on the fixed address. If you enter this command when an LECS is not a master, the LECS will listen on this address when it becomes a master. If you do not enter this command, the LECS will not listen on the fixed address.

Multiple commands that assign ATM addresses to the LECS can be issued on the same interface in order to assign different ATM addresses to the LECS. Commands that assign ATM addresses to the LECS include **lane auto-config-atm-address, lane config-atm-address**, and **lane fixed-config-atm-address**. The **lane config database** command and at least one command that assigns an ATM address to the LECS are required to activate a LECS.

Examples

The following example associates the LANE configuration server with the database named *network1* and specifies that the configuration server's ATM address is the fixed address:

```
lane database network1
 name eng server-atm-address 39.020304050607080910111213.0800.AA00.1001.02
 name mkt server-atm-address 39.020304050607080910111213.0800.AA00.4001.01
```

```
interface atm 1/0
 lane config database network1
 lane config fixed-config-atm-address
```

The following example causes the LANE server and LANE client on the subinterface to use the fixed ATM address to communicate with the configuration server:

```
interface atm 2/0.1
 ip address 172.16.0.4 255.255.255.0
 lane client ethernet
 lane server-bus ethernet eng
 lane fixed-config-atm-address
```

Related Commands

You can use the master index or search online to find documentation of related commands.

lane auto-config-atm-address
lane config-atm-address
lane config database

LANE GLOBAL-LECS-ADDRESS

To specify a list of LECS addresses to use when the addresses cannot be obtained from the ILMI, use the **lane global-lecs-address** interface configuration command. The **no** form of this command removes an LECS address from the list.

lane global-lecs-address *address*
no lane global-lecs-address *address*

Syntax	*Description*
address	Address of the LECS. You cannot enter the well-known LECS address.

Default

No addresses are configured. The router obtains LECS addresses from the ILMI.

Command Mode

Interface configuration

Usage Guidelines

This command first appeared in Cisco IOS Release 11.2.

Use this command when your ATM switches do not support the ILMI list of LECS addresses and you want to configure Simple Server Redundancy. This command will simulate the list of LECS addresses, as if they were obtained from the ILMI. Enter this command with a different address for each LECS. The order they are entered determines their priority. You should enter the addresses in the same order as you would on the ATM switch.

NOTES

You must configure the same list of addresses on each interface that contains a LANE entity.

If your switches do support ILMI, this command forces the router to use the addresses specified and will not use the ILMI to obtain the LECS addresses.

Since the well-known LECS address is always used as a last resort LECS address, you cannot use the address in this command.

LANE LE-ARP

To add a static entry to the LE ARP table of the LANE client configured on the specified subinterface, use the **lane le-arp** interface configuration command. To remove a static entry from the LE ARP table of the LANE client on the specified subinterface, use the **no** form of this command.

lane le-arp {*mac-address* | **route-desc segment** *segment-number* **bridge** *bridge-number*} *atm-address*

no lane le-arp {*mac-address* | **route-desc segment** *segment-number* **bridge** *bridge-number*} *atm-address*

Syntax	*Description*
mac-address	MAC address to bind to the specified ATM address.
segment-number	LANE segment number. The segment number ranges from 1 to 4095.

Syntax	*Description*
bridge-number	Bridge number that is contained in the route descriptor. The bridge number ranges from 1 to 15.
atm-address	ATM address.

Default

No static address bindings are provided.

Command Mode

Interface configuration

Usage Guidelines

This command first appeared in Cisco IOS Release 11.0.

This command adds or removes a static entry binding a MAC address or segment number and bridge number to an ATM address. It does not add or remove dynamic entries. Removing the static entry for a specified ATM address from an LE ARP table does not release Data Direct VCCs established to that ATM address. However, clearing a static entry clears any fast-cache entries that were created from the MAC address-to-ATM address binding.

Static LE ARP entries are neither aged nor removed automatically.

To remove dynamic entries from the LE ARP table of the LANE client on the specified sub-interface, use the **clear lane le-arp** command.

Example

The following example adds a static entry to the LE ARP table:

```
lane le-arp 0800.aa00.0101 47.000014155551212f.00.00.0800.200C.1001.01
```

The following example adds a static entry to the LE ARP table binding segment number 1, bridge number 1 to the ATM address:

```
lane le-arp route-desc segment 1 bridge 1 39.020304050607080910111213.00000CA05B41.01
```

Related Commands

You can use the master index or search online to find documentation of related commands.

clear lane le-arp

LANE SERVER-ATM-ADDRESS

To specify an ATM address, and thus override the automatic ATM address assignment, for the LANE server on the specified subinterface, use the **lane server-atm-address** interface configuration command. To remove the ATM address previously specified for the LANE server on the specified subinterface and thus revert to the automatic address assignment, use the **no** form of this command.

lane server-atm-address *atm-address-template*
no server-atm-address [*atm-address-template*]

Syntax	*Description*
atm-address-template	ATM address or a template in which wildcard characters are replaced by any nibble or group of nibbles of the prefix bytes, the ESI bytes, or the selector byte of the automatically assigned ATM address.

Defaults

For the LANE server, the default is automatic address assignment; the LANE client finds the LANE server by consulting the configuration server.

Command Mode

Interface configuration

Usage Guidelines

This command first appeared in Cisco IOS Release 11.0.

This command also instructs the LANE client configured on this subinterface to reach the LANE server by using the specified ATM address instead of the ATM address provided by the configuration server.

When used on a selected subinterface, but with a different ATM address than was used previously, this command replaces the LANE server's ATM address.

ATM Addresses. A LANE ATM address has the same syntax as an NSAP (but it is not a network-level address). It consists of the following:

- A 13-byte prefix that includes the following fields defined by the ATM Forum:
  - AFI (Authority and Format Identifier) field (1 byte)
  - DCC (Data Country Code) or ICD (International Code Designator) field (2 bytes)
  - DFI field (Domain Specific Part Format Identifier) (1 byte)
  - Administrative Authority field (3 bytes)
  - Reserved field (2 bytes)
  - Routing Domain field (2 bytes)
  - Area field (2 bytes)
- A 6-byte end-system identifier (ESI)
- A 1-byte selector field

Address Templates. LANE ATM address templates can use two types of wildcards: an asterisk (*) to match any single character (nibble), and an ellipsis (...) to match any number of leading, middle, or trailing characters. The values of the characters replaced by wildcards come from the automatically assigned ATM address.

In LANE, a *prefix template* explicitly matches the prefix, but uses wildcards for the ESI and selector fields. An *ESI template* explicitly matches the ESI field, but uses wildcards for the prefix and selector.

In our implementation of LANE, the prefix corresponds to the switch, the ESI corresponds to the ATM interface, and the Selector field corresponds to the specific subinterface of the interface.

Examples

The following example uses an ESI template to specify the part of the ATM address corresponding to the interface; the remaining parts of the ATM address come from automatic assignment:

```
lane server-atm-address ...0800.200C.1001.**
```

The following example uses a prefix template to specify the part of the ATM address corresponding to the switch; the remaining part of the ATM address come from automatic assignment:

```
lane server-atm-address 45.000014155551212f.00.00...
```

Related Commands

You can use the master index or search online to find documentation of related commands.

lane server-bus

LANE SERVER-BUS

To enable a LANE server and a broadcast-and-unknown server on the specified subinterface, use the **lane server-bus** interface configuration command. To disable a LANE server and broadcast-and-unknown server on the specified subinterface, use the **no** form of this command.

lane server-bus {ethernet | tokenring} *elan-name*
no lane server-bus [{ethernet | tokenring} *elan-name*]

Syntax	*Description*
ethernet	Identifies the emulated LAN attached to this subinterface as an Ethernet ELAN.
tokenring	Identifies the emulated LAN attached to this subinterface as a Token Ring ELAN.
elan-name	Name of the emulated LAN. The maximum length of the name is 32 characters.

Defaults

No LAN type or emulated LAN name is provided.

Command Mode

Interface configuration

Usage Guidelines

This command first appeared in Cisco IOS Release 11.0.

This command is ordinarily used.

The LANE server and the broadcast-and-unknown server (BUS) are located on the same router.

If a **lane server-bus** command has already been entered on the subinterface for a different emulated LAN, then the server initiates termination procedures with all clients and comes up as the server for the new emulated LAN.

The **no** form of this command removes a previously configured LANE server and broadcast-and-unknown server on the subinterface.

Example

The following example enables a LANE server and broadcast-and unknown server for a Token Ring ELAN:

```
lane server-bus tokenring
```

Related Commands

You can use the master index or search online to find documentation of related commands.

lane server-atm-address

NAME LOCAL-SEG-ID

To specify or replace the ring number of the emulated LAN in the configuration server's configuration database, use the **name local-seg-id** database configuration command. To remove the ring number from the database, use the **no** form of this command.

name *elan-name* **local-seg-id** *segment-number*
no name *elan-name* **local-seg-id** *segment-number*

Syntax	*Description*
elan-name	Name of the emulated LAN. The maximum length of the name is 32 characters.
segment-number	Segment number to be assigned to the emulated LAN. The number ranges from 1 to 4095.

Default

No emulated LAN name or segment number is provided.

Command Mode

Database configuration

Usage Guidelines

This command first appeared in Cisco IOS Release 11.3.

This command is ordinarily used for Token Ring LANE.

The same LANE ring number cannot be assigned to more than one emulated LAN.

The **no** form of this command deletes the relationships.

Example

The following example specifies a ring number of 1024 for the emulated LAN *red*:

```
name red local-seg-id 1024
```

Related Commands

You can use the master index or search online to find documentation of related commands.

default-name
lane database
mac-address name

NAME SERVER-ATM-ADDRESS

To specify or replace the ATM address of the LANE server for the emulated LAN in the configuration server's configuration database, use the **name server-atm-address** database configuration command. To remove it from the database, use the **no** form of this command.

name *elan-name* **server-atm-address** *atm-address* [**restricted** | **un-restricted**] [**index** *number*]

no name *elan-name* **server-atm-address** *atm-address* [**restricted** | **un-restricted**] [**index** *number*]

Syntax	*Description*
elan-name	Name of the emulated LAN. Maximum length is 32 characters.
atm-address	LANE server's ATM address.
restricted | **unrestricted**	(Optional) Membership in the named emulated LAN is restricted to the LANE clients explicitly defined to the emulated LAN in the configuration server's database.
index *number*	(Optional) Priority number. When specifying multiple LANE servers for fault tolerance, you can specify a priority for each server; 0 is the highest priority.

Defaults

No emulated LAN name or server ATM address are provided.

Command Mode

Database configuration

Usage Guidelines

This command first appeared in Cisco IOS Release 11.0. The **restricted** command first appeared in Cisco IOS Release 11.1. The **unrestricted** and **index** keywords first appeared in Cisco IOS Release 11.2. Emulated LAN names must be unique within one named LANE configuration database.

Specifying an existing emulated LAN name with a new LANE server ATM address adds the LANE server ATM address for that emulated LAN for redundant server operation or simple LANE service replication. This command can be entered multiple times.

The **no** form of this command deletes the relationships.

Example

The following example configures the *example3* database with two restricted and one unrestricted emulated LANs. The clients that can be assigned to the *eng* and *mkt* emulated LANs are specified using the **client-atm-address** commands. All other clients are assigned to the *man* emulated LAN.

```
lane database example3
 name eng server-atm-address 39.000001415555121101020304.0800.200c.1001.02 restricted
 name man server-atm-address 39.000001415555121101020304.0800.200c.1001.01
 name mkt server-atm-address 39.000001415555121101020304.0800.200c.4001.01 restricted
 client-atm-address 39.000001415555121101020304.0800.200c.1000.02 name eng
 client-atm-address 39.0000001415555121101020304.0800.200c.2000.02 name eng
 client-atm-address 39.000001415555121101020304.0800.200c.3000.02 name mkt
 client-atm-address 39.000001415555121101020304.0800.200c.4000.01 name mkt
 default-name man
```

Related Commands

You can use the master index or search online to find documentation of related commands.

client-atm-address name
default-name
lane database
mac-address name

SHOW LANE

To display detailed information for all the LANE components configured on an interface or any of its subinterfaces, on a specified subinterface, or on an emulated LAN, use the **show lane** EXEC command.

> **show lane [interface atm** *slot/port*[*.subinterface-number*] | **name** *elan-name*] **[brief]**
> (for the AIP on the Cisco 7500 series routers;
> for the ATM port adapter on the Cisco 7200 series)

show lane [**interface atm** *slot/port-adapter/port*[*.subinterface-number*] | **name** *elan-name*] [**brief**] (for the ATM port adapter on the Cisco 7500 series routers)
show lane [**interface atm** *number*[*.subinterface-number*] | **name** *elan-name*] [**brief**] (for the Cisco 4500 and 4700 routers)

Syntax	*Description*
interface atm *slot/port*	(Optional) ATM interface slot and port for the following: AIP on the Cisco 7500 series routers. ATM port adapter on the Cisco 7200 series routers.
interface atm *slot/port-adapter/port*	(Optional) ATM interface slot, port adapter, and port number for the ATM port adapter on the Cisco 7500 series routers.
interface atm *number*	(Optional) ATM interface number for the NPM on the Cisco 4500 or 4700 routers.
.subinterface-number	(Optional) Subinterface number.
name *elan-name*	(Optional) Name of emulated LAN. The maximum length of the name is 32 characters.
brief	(Optional) Keyword used to display the brief subset of available information.

Command Mode

EXEC

Usage Guidelines

This command first appeared in Cisco IOS Release 11.0.

Entering the **show lane** command is equivalent to entering the **show lane config**, **show lane server**, **show lane bus**, and **show lane client** commands. The **show lane** command shows all LANE-related information except the **show lane database** command information.

Sample Displays

The following is sample output of the **show lane** command for an Ethernet-emulated LAN:

```
Router# show lane

LE Config Server ATM2/0 config table: cisco_eng
```

```
Admin: up  State: operational
LECS Mastership State: active master
list of global LECS addresses (30 seconds to update):
39.020304050607080910111213.00000CA05B43.00  <-------- me
ATM Address of this LECS: 39.020304050607080910111213.00000CA05B43.00 (auto)
 vcd  rxCnt txCnt  callingParty
  50      2      2  39.020304050607080910111213.00000CA05B41.02 LES elan2 0 active
cumulative total number of unrecognized packets received so far: 0
cumulative total number of config requests received so far: 30
cumulative total number of config failures so far: 12
    cause of last failure: no configuration
    culprit for the last failure: 39.020304050607080910111213.00602F557940.01

LE Server ATM2/0.2  ELAN name: elan2  Admin: up  State: operational
type: ethernet          Max Frame Size: 1516
ATM address: 39.020304050607080910111213.00000CA05B41.02
LECS used: 39.020304050607080910111213.00000CA05B43.00 connected, vcd 51
control distribute: vcd 57, 2 members, 2 packets

proxy/ (ST: Init, Conn, Waiting, Adding, Joined, Operational, Reject, Term)
lecid ST vcd    pkts Hardware Addr  ATM Address
   1  O   54       2 0000.0ca0.5b40 39.020304050607080910111213.00000CA05B40.02
   2  O   81       2 0060.2f55.7940 39.020304050607080910111213.00602F557940.02

LE BUS ATM2/0.2  ELAN name: elan2  Admin: up  State: operational
type: ethernet          Max Frame Size: 1516
ATM address: 39.020304050607080910111213.00000CA05B42.02
data forward: vcd 61, 2 members, 0 packets, 0 unicasts

lecid  vcd     pkts   ATM Address
    1   58        0 39.020304050607080910111213.00000CA05B40.02
    2   82        0 39.020304050607080910111213.00602F557940.02

LE Client ATM2/0.2  ELAN name: elan2  Admin: up  State: operational
Client ID: 1                 LEC up for 11 minutes 49 seconds
Join Attempt: 1
HW Address: 0000.0ca0.5b40   Type: ethernet             Max Frame Size: 1516

ATM Address: 39.020304050607080910111213.00000CA05B40.02

 VCD  rxFrames  txFrames  Type       ATM Address
   0         0         0  configure  39.020304050607080910111213.00000CA05B43.00
  55         1         4  direct     39.020304050607080910111213.00000CA05B41.02
  56         6         0  distribute 39.020304050607080910111213.00000CA05B41.02
  59         0         1  send       39.020304050607080910111213.00000CA05B42.02
  60         3         0  forward    39.020304050607080910111213.00000CA05B42.02
  84         3         5  data       39.020304050607080910111213.00602F557940.02
```

The following is sample output of the **show lane** command for a Token Ring LANE network:

```
Router# show lane
```

```
LE Config Server ATM4/0 config table: eng
Admin: up  State: operational
LECS Mastership State: active master
list of global LECS addresses (35 seconds to update):
39.020304050607080910111213.006047704183.00  <-------- me
ATM Address of this LECS: 39.020304050607080910111213.006047704183.00 (auto)
 vcd  rxCnt  txCnt  callingParty
   7      1      1  39.020304050607080910111213.006047704181.01 LES elan1 0 active
cumulative total number of unrecognized packets received so far: 0
cumulative total number of config requests received so far: 2
cumulative total number of config failures so far: 0

LE Server ATM4/0.1  ELAN name: elan1  Admin: up  State: operational
type: token ring         Max Frame Size: 4544      Segment ID: 2048
ATM address: 39.020304050607080910111213.006047704181.01
LECS used: 39.020304050607080910111213.006047704183.00 connected, vcd 9
control distribute: vcd 12, 1 members, 2 packets

proxy/ (ST: Init, Conn, Waiting, Adding, Joined, Operational, Reject, Term)
lecid ST vcd    pkts Hardware Addr  ATM Address
   1  O   8       3 100.2          39.020304050607080910111213.006047704180.01
                    0060.4770.4180 39.020304050607080910111213.006047704180.01

LE BUS ATM4/0.1  ELAN name: elan1  Admin: up  State: operational
type: token ring         Max Frame Size: 4544      Segment ID: 2048
ATM address: 39.020304050607080910111213.006047704182.01
data forward: vcd 16, 1 members, 0 packets, 0 unicasts

lecid  vcd     pkts   ATM Address
    1   13        0 39.020304050607080910111213.006047704180.01

LE Client ATM4/0.1  ELAN name: elan1  Admin: up  State: operational
Client ID: 1                  LEC up for 2 hours 25 minutes 39 seconds
Join Attempt: 3
HW Address: 0060.4770.4180   Type: token ring          Max Frame Size: 4544
Ring:100    Bridge:2        ELAN Segment ID: 2048
ATM Address: 39.020304050607080910111213.006047704180.01

 VCD  rxFrames  txFrames  Type        ATM Address
   0         0         0  configure   39.020304050607080910111213.006047704183.00
  10         1         3  direct      39.020304050607080910111213.006047704181.01
  11         2         0  distribute  39.020304050607080910111213.006047704181.01
  14         0         0  send        39.020304050607080910111213.006047704182.01
  15         0         0  forward     39.020304050607080910111213.006047704182.01
```

Table 9–1 describes significant fields in the sample displays.

Table 9–1 *Show LANE Field Descriptions*

Field	Description
LE Config Server	Identifies the following lines as applying to the LANE configuration server. These lines are also displayed in output from the **show lane config** command. See the **show lane config** command for explanations of the output.
LE Server	Identifies the following lines as applying to the LANE server. These lines are also displayed in output from the **show lane server** command. See the **show lane server** command for explanations of the output.
LE BUS	Identifies the following lines as applying to the LANE broadcast-and-unknown server. These lines are also displayed in output from the **show lane bus** command. See the **show lane bus** command for explanations of the output.
LE Client	Identifies the following lines as applying to a LANE client. These lines are also displayed in output from the **show lane client** command. See the **show lane bus** command for explanations of the output.

SHOW LANE BUS

To display detailed LANE information for the broadcast-and-unknown server configured on an interface or any of its subinterfaces, on a specified subinterface, or on an emulated LAN, use the **show lane bus** EXEC command:

show lane bus [**interface atm** *slot/port*[*.subinterface-number*] | **name** *elan-name*] [**brief**] (for the AIP on the Cisco 7500 series routers; for the ATM port adapter on the Cisco 7200 series)

show lane bus [**interface atm** *slot/port-adapter/port*[*.subinterface-number*] | **name** *elan-name*] [**brief**] (for the ATM port adapter on the Cisco 7500 series routers)

show lane bus [**interface atm** *number*[*.subinterface-number*] | **name** *elan-name*] [**brief**] (for the Cisco 4500 and 4700 routers)

Syntax	*Description*
interface atm *slot/port*	(Optional) ATM interface slot and port for the following: • AIP on the Cisco 7500 series routers. • ATM port adapter on the Cisco 7200 series routers.

Syntax	*Description*
interface atm *slot/port-adapter/port*	(Optional) ATM interface slot, port adapter, and port number for the ATM port adapter on the Cisco 7500 series routers.
interface atm *number*	(Optional) ATM interface number for the NPM on the Cisco 4500 or 4700 routers.
.subinterface-number	(Optional) Subinterface number.
name *elan-name*	(Optional) Name of emulated LAN. The maximum length of the name is 32 characters.
brief	(Optional) Keyword used to display the brief subset of available information.

Command Mode

EXEC

Usage Guidelines

This command first appeared in Cisco IOS Release 11.0.

Sample Displays

The following is sample output of the **show lane bus** command for an Ethernet-emulated LAN:

```
Router# show lane bus
LE BUS ATM2/0.2  ELAN name: elan2  Admin: up  State: operational
type: ethernet          Max Frame Size: 1516
ATM address: 39.020304050607080910111213.00000CA05B42.02
data forward: vcd 61, 2 members, 0 packets, 0 unicasts
lecid  vcd     pkts   ATM Address
    1   58        0 39.020304050607080910111213.00000CA05B40.02
    2   82        0 39.020304050607080910111213.00602F557940.02
```

The following is sample output of the **show lane bus** command for a Token Ring LANE:

```
Router# show lane bus
LE BUS ATM3/0.1  ELAN name: anubis  Admin: up  State: operational
type: token ring          Max Frame Size: 4544      Segment ID: 2500
ATM address: 47.00918100000000000000000.00000CA01662.01
data forward: vcd 14, 2 members, 0 packets, 0 unicasts
lecid  vcd     pkts   ATM Address
    1   11        0 47.00918100000000000000000.00000CA01660.01
    2   17        0 47.00918100000000000000000.00000CA04960.01
```

Table 9–2 describes significant fields in the sample displays.

Table 9–2 *Show LANE BUS Field Descriptions*

Field	Description
LE BUS ATM2/0.2	Interface and subinterface for which information is displayed.
ELAN name	Name of the emulated LAN for this broadcast-and-unknown server.
Admin	Administrative state, either up or down.
State	Status of this LANE broadcast-and-unknown server. Possible states include down and operational.
type	Type of emulated LAN.
Max Frame Size	Maximum frame size (in bytes) on the emulated LAN.
Segment ID	The emulated LAN's ring number. This field appears only for Token Ring LANE.
ATM address	ATM address of this LANE broadcast-and-unknown server.
data forward	Virtual channel descriptor of the Data Forward VCC, the number of LANE clients attached to the VCC, and the number of packets transmitted on the VCC.
lecid	Identifier assigned to each LANE client on the Data Forward VCC.
vcd	Virtual channel descriptor used to reach the LANE client.
pkts	Number of packets sent by the broadcast-and-unknown server to the LANE client.
ATM Address	ATM address of the LANE client.

SHOW LANE CLIENT

To display detailed LANE information for all the LANE clients configured on an interface or any of its subinterfaces, on a specified subinterface, or on an emulated LAN, use the **show lane client** EXEC command.

show lane client [**interface atm** *slot/port*[.*subinterface-number*] | **name** *elan-name*] [**brief**] (for the AIP on the Cisco 7500 series routers; for the ATM port adapter on the Cisco 7200 series)
show lane client [**interface atm** *slot/port-adapter/port*[.*subinterface-number*] | **name** *elan-name*] [**brief**] (for the ATM port adapter on the Cisco 7500 series routers)
show lane client [**interface atm** *number*[.*subinterface-number*] | **name** *elan-name*] [**brief**] (for the Cisco 4500 and 4700 routers)

Syntax	*Description*
interface atm *slot/port*	(Optional) ATM interface slot and port for the following: • AIP on the Cisco 7500 series routers. • ATM port adapter on the Cisco 7200 series routers.
interface atm *slot/port-adapter/port*	(Optional) ATM interface slot, port adapter, and port number for the ATM port adapter on the Cisco 7500 series routers.
interface atm *number*	(Optional) ATM interface number for the NPM on the Cisco 4500 or 4700 routers.
.subinterface-number	(Optional) Subinterface number.
name *elan-name*	(Optional) Name of emulated LAN. The maximum length of the name is 32 characters.
brief	(Optional) Keyword used to display the brief subset of available information.

Command Mode

EXEC

Usage Guidelines

This command first appeared in Cisco IOS Release 11.0.

Sample Displays

The following is sample output from the **show lane client** command for an Ethernet-emulated LAN:

```
Router# show lane client
LE Client ATM2/0.2  ELAN name: elan2  Admin: up  State: operational
Client ID: 1                 LEC up for 11 minutes 49 seconds
Join Attempt: 1
```

```
HW Address: 0000.0ca0.5b40   Type: ethernet            Max Frame Size: 1516

ATM Address: 39.020304050607080910111213.00000CA05B40.02

 VCD  rxFrames  txFrames  Type        ATM Address
   0         0         0  configure   39.020304050607080910111213.00000CA05B43.00
  55         1         4  direct      39.020304050607080910111213.00000CA05B41.02
  56         6         0  distribute  39.020304050607080910111213.00000CA05B41.02
  59         0         1  send        39.020304050607080910111213.00000CA05B42.02
  60         3         0  forward     39.020304050607080910111213.00000CA05B42.02
  84         3         5  data        39.020304050607080910111213.00602F557940.02
```

The following is sample output from the **show lane client** command for a Token Ring LANE:

```
Router# show lane client

LE Client ATM4/0.1  ELAN name: elan1  Admin: up  State: operational
Client ID: 1                   LEC up for 2 hours 26 minutes 3 seconds
Join Attempt: 3
HW Address: 0060.4770.4180   Type: token ring            Max Frame Size: 4544
Ring:100    Bridge:2        ELAN Segment ID: 2048
ATM Address: 39.020304050607080910111213.006047704180.01

 VCD  rxFrames  txFrames  Type        ATM Address
   0         0         0  configure   39.020304050607080910111213.006047704183.00
  10         1         3  direct      39.020304050607080910111213.006047704181.01
  11         2         0  distribute  39.020304050607080910111213.006047704181.01
  14         0         0  send        39.020304050607080910111213.006047704182.01
  15         0         0  forward     39.020304050607080910111213.006047704182.01
```

Table 9–3 describes significant fields in the sample displays.

Table 9–3 *Show LANE Client Field Descriptions*

Field	Description
LE Client ATM2/0.2	Interface and subinterface of this client.
ELAN name	Name of the emulated LAN.
Admin	Administrative state; either up or down.
State	Status of this LANE client. Possible states include initialState, lecsConnect, configure, join, busConnect, and operational.
Client ID	The LAN emulation 2-byte Client ID assigned by the LAN emulation server.
Join Attempt	The number of attempts before successfully joining the emulated LAN.

Table 9–3 *Show LANE Client Field Descriptions, Continued*

Field	Description
HW Address	MAC address of this LANE client.
Type	Type of emulated LAN.
Max Frame Size	Maximum frame size (in bytes) on the emulated LAN.
Ring	The ring number for the client. This field only appears for Token Ring LANE.
Bridge	The bridge number for the client. This field only appears for Token Ring LANE.
ELAN Segment ID	The ring number for the emulated LAN. This field only appears for Token Ring LANE.
ATM Address	ATM address of this LANE client.
VCD	Virtual channel descriptor for each of the VCCs established for this LANE client.
rxFrames	Number of frames received.
txFrames	Number of frames transmitted.
Type	Type of VCC. The Configure Direct VCC is shown in this display as *configure*. The Control Direct VCC is shown as *direct*; the Control Distribute VCC is shown as *distribute*. The Multicast Send VCC and Multicast Forward VC are shown as *send* and *forward*, respectively. The Data Direct VCC is shown as *data*
ATM Address	ATM address of the LANE component at the other end of this VCC.

SHOW LANE CONFIG

To display global LANE information for the configuration server configured on an interface, use the **show lane config** EXEC command.

show lane config [interface atm *slot*/0] (for the AIP on the Cisco 7500 series routers; for the ATM port adapter on the Cisco 7200 series)
show lane config [interface atm *slot*/*port-adapter*/0] (for the ATM port adapter on the Cisco 7500 series routers)
show lane config [interface atm *number*] (for the Cisco 4500 and 4700 routers)

Syntax	*Description*
interface atm *slot*/0	(Optional) ATM interface slot and port for the following: • AIP on the Cisco 7500 series routers. • ATM port adapter on the Cisco 7200 series routers.
interface atm *slot/port-adapter*/0	(Optional) ATM interface slot, port adapter, and port number for the ATM port adapter on the Cisco 7500 series routers.
interface atm *number*	(Optional) ATM interface number for the NPM on the Cisco 4500 or 4700 routers.

Command Mode

EXEC

Usage Guidelines

This command first appeared in Cisco IOS Release 11.0.

Sample Displays

The following is sample **show lane config** output for an Ethernet emulated LAN:

```
Router# show lane config

LE Config Server ATM2/0 config table: cisco_eng
Admin: up  State: operational
LECS Mastership State: active master
list of global LECS addresses (30 seconds to update):
39.020304050607080910111213.00000CA05B43.00  <-------- me
ATM Address of this LECS: 39.020304050607080910111213.00000CA05B43.00 (auto)
 vcd  rxCnt  txCnt  callingParty
  50      2      2  39.020304050607080910111213.00000CA05B41.02 LES elan2 0 active
cumulative total number of unrecognized packets received so far: 0
cumulative total number of config requests received so far: 30
cumulative total number of config failures so far: 12
    cause of last failure: no configuration
    culprit for the last failure: 39.020304050607080910111213.00602F557940.01
```

The following example shows sample **show lane config** output for TR-LANE:

```
Router# show lane config

LE Config Server ATM4/0 config table: eng
Admin: up  State: operational
LECS Mastership State: active master
list of global LECS addresses (40 seconds to update):
39.020304050607080910111213.006047704183.00  <-------- me
ATM Address of this LECS: 39.020304050607080910111213.006047704183.00 (auto)
```

```
  vcd  rxCnt  txCnt  callingParty
    7      1      1  39.020304050607080910111213.006047704181.01 LES elan1 0 active
 cumulative total number of unrecognized packets received so far: 0
 cumulative total number of config requests received so far: 2
 cumulative total number of config failures so far: 0
```

Table 9–4 describes significant fields in the sample display.

Table 9–4 *Show LANE Config Command Field Descriptions*

Field	**Description**
LE Config Server	Major interface on which the LANE configuration server is configured.
config table	Name of the database associated with the LANE configuration server.
Admin	Administrative state, either up or down.
State	State of the configuration server: down or operational. If down, the reasons field indicates why it is down. The reasons include the following: NO-config-table, NO-nsap-address, and NO-interface-up.
LECS Mastership state	Mastership state of the configuration server. If you have configured simple server redundancy, the configuration server with the lowest index is the active LECS.
list of global LECS addresses	List of LECS addresses.
40 seconds to update	Amount of time until the next update.
<-------- me	ATM address of this configuration server.
ATM Address of this LECS	ATM address of the active configuration server.
auto	Method of ATM address assignment for the configuration server. In this example, the address is assigned by the automatic method.
vcd	Virtual circuit descriptor that uniquely identifies the configure VCC.
rxCnt	Number of packets received.

Table 9–4 *Show LANE Config Command Field Descriptions, Continued*

Field	Description
txCnt	Number of packets transmitted.
callingParty	ATM NSAP address of the LANE component that is connected to the LECS. 'elan1' indicates the emulated LAN name, '0' indicates the priority number, and 'active' indicates that the server is active.

SHOW LANE DATABASE

To display the configuration server's database, use the **show lane database** EXEC command.

show lane database [*database-name*]

Syntax	*Description*
database-name	(Optional) Specific database name.

Command Mode

EXEC

Usage Guidelines

This command first appeared in Cisco IOS Release 11.0.

By default, this command displays the LANE configuration server information displayed by the **show lane config** command.

If no database name is specified, this command shows all databases.

Sample Displays

The following is sample output of the **show lane database** command for an Ethernet LANE:

```
Router# show lane database

LANE Config Server database table 'engandmkt' bound to interface/s: ATM1/0
default elan: none
elan 'eng': restricted
  server 45.000001415555121f.yyyy.zzzz.0800.200c.1001.01 (prio 0) active
  LEC MAC  0800.200c.1100
```

```
    LEC NSAP 45.000001415555121f.yyyy.zzzz.0800.200c.1000.01
    LEC NSAP 45.000001415555124f.yyyy.zzzz.0800.200c.1300.01
  elan 'mkt':
    server 45.000001415555121f.yyyy.zzzz.0800.200c.1001.02 (prio 0) active
    LEC MAC  0800.200c.1200
    LEC NSAP 45.000001415555121f.yyyy.zzzz.0800.200c.1000.02
    LEC NSAP 45.000001415555124f.yyyy.zzzz.0800.200c.1300.02
```

The following is sample output of the **show lane database** command for a Token Ring LANE:

```
Router# show lane database

LANE Config Server database table 'eng' bound to interface/s: ATM4/0
default elan: elan1
elan 'elan1': un-restricted, local-segment-id 2048
  server 39.020304050607080910111213.006047704181.01 (prio 0) active
```

Table 9–5 describes significant fields in the sample displays.

Table 9–5 *Show LANE Database Command Field Descriptions*

Field	Description
LANE Config Server database	Name of this database and interfaces bound to it.
default elan	Default name, if one is established.
elan	Name of the emulated LAN whose data is reported in this line and the following indented lines.
unrestricted	Indicates whether this emulated LAN is restricted or unrestricted.
local-segment-id 2048	Ring number of the emulated LAN.
server	ATM address of the configuration server.
(prio 0) active	Priority level and simple server redundancy state of this configuration server. If you have configured simple server redundancy, the configuration server with the lowest priority will be active.
LEC MAC	MAC addresses of an individual LANE client in this emulated LAN. This display includes a separate line for every LANE client in this emulated LAN.
LEC NSAP	ATM addresses of all LANE clients in this emulated LAN.

SHOW LANE DEFAULT-ATM-ADDRESSES

To display the automatically assigned ATM address of each LANE component in a router or on a specified interface or subinterface, use the **show lane default-atm-addresses** EXEC command.

show lane default-atm-addresses [**interface atm** *slot/port.subinterface-number*] (for the AIP on the Cisco 7500 series routers; for the ATM port adapter on the Cisco 7200 series)

show lane default-atm-addresses [**interface atm** *slot/port-adapter/port.subinterface-number*] (for the ATM port adapter on the Cisco 7500 series routers)

show lane default-atm-addresses [**interface atm** *number.subinterface-number*] (for the Cisco 4500 and 4700 routers)

Syntax	*Description*
interface atm *slot/port*	(Optional) ATM interface slot and port for the following: • AIP on the Cisco 7500 series routers. • ATM port adapter on the Cisco 7200 series routers.
interface atm *slot/port-adapter/port*	(Optional) ATM interface slot, port adapter, and port number for the ATM port adapter on the Cisco 7500 series routers.
interface atm *number*	(Optional) ATM interface number for the NPM on the Cisco 4500 or 4700 routers.
.subinterface-number	(Optional) Subinterface number.

Command Mode

EXEC

Usage Guidelines

The **show lane default-atm-addresses** [**interface atm** *slot/port.subinterface-number*] command first appeared in Cisco IOS Release 11.0.

The **show lane default-atm-addresses** [**interface atm** *number.subinterface-number*] command first appeared in Cisco IOS Release 11.1.

It is not necessary to have any of the LANE components running on this router before you use this command.

Sample Display

The following is sample output of the **show lane default-atm-addresses** command for the ATM interface 1/0 when all the major LANE components are located on that interface:

```
Router# show lane default-atm-addresses interface atm1/0

interface ATM1/0:
LANE Client:        47.00000000000000000000000.00000C304A98.**
LANE Server:        47.00000000000000000000000.00000C304A99.**
LANE Bus:           47.00000000000000000000000.00000C304A9A.**
LANE Config Server: 47.00000000000000000000000.00000C304A9B.00
note: ** is the subinterface number byte in hex
```

Table 9–6 describes significant fields shown in the display.

Table 9–6 *Show LANE Default-ATM-Addresses Field Descriptions*

Field	Description
interface ATM1/0:	Specified interface.
LANE Client:	ATM address of the LANE client on the interface.
LANE Server:	ATM address of the LANE server on the interface.
LANE Bus:	ATM address of the LANE broadcast-and-unknown server on the interface.
LANE Config Server:	ATM address of the LANE configuration server on the interface.

SHOW LANE LE-ARP

To display the LANE ARP table of the LANE client configured on an interface or any of its subinterfaces, on a specified subinterface, or on an emulated LAN, use the **show lane le-arp** EXEC command.

show lane le-arp [interface atm *slot/port*[*.subinterface-number*] | **name** *elan-name*] (for the AIP on the Cisco 7500 series routers; for the ATM port adapter on the Cisco 7200 series)

show lane le-arp [interface atm *slot/port-adapter/port*[*.subinterface-number*] | **name** *elan-name*] (for the ATM port adapter on the Cisco 7500 series routers)

show lane le-arp [interface atm *number*[*.subinterface-number*] | **name** *elan-name*] (for the Cisco 4500 and 4700 routers)

Syntax	*Description*
interface atm *slot/port*	(Optional) ATM interface slot and port for the following: • AIP on the Cisco 7500 series routers. • ATM port adapter on the Cisco 7200 series routers.
interface atm *slot/port-adapter/port*	(Optional) ATM interface slot, port adapter, and port number for the ATM port adapter on the Cisco 7500 series routers.
interface atm *number*	(Optional) ATM interface number for the NPM on the Cisco 4500 or 4700 routers.
.subinterface-number	(Optional) Subinterface number.
name *elan-name*	(Optional) Name of emulated LAN. The maximum length of the name is 32 characters.

Command Mode

EXEC

Usage Guidelines

This command first appeared in Cisco IOS Release 11.0.

Sample Displays

The following is sample output of the **show lane le-arp** command for an Ethernet LANE client:

```
Router# show lane le-arp

Hardware Addr   ATM Address                                        VCD  Interface
0000.0c15.a2b5  39.000000000000000000000000.00000C15A2B5.01  39   ATM1/0.1
0000.0c15.f3e5  39.000000000000000000000000.00000C15F3E5.01  25*  ATM1/0.1
```

The following is sample output of the **show lane le-arp** command for a Token Ring LANE client:

```
Router# show lane le-arp

Ring Bridge       ATM Address                                   VCD  Interface
512   6           39.020304050607080910111213.00602F557940.01  47   ATM2/0.1
```

Table 9–7 describes significant fields shown in the displays.

Table 9–7 *Show LANE LE-ARP Field Descriptions*

Field	Description
Hardware Addr	MAC address, in dotted hexadecimal notation, assigned to the LANE component at the other end of this VCD.
Ring	Route descriptor segment number for the LANE component.
Bridge	Bridge number for the LANE component.
ATM Address	ATM address of the LANE component at the other end of this VCD.
VCD	Virtual circuit descriptor.
Interface	Interface or subinterface used to reach the specified component.

SHOW LANE SERVER

To display global information for the LANE server configured on an interface, on any of its subinterfaces, on a specified subinterface, or on an emulated LAN, use the **show lane server** EXEC command.

show lane server [**interface atm** *slot/port*[*.subinterface-number*] | **name** *elan-name*] [**brief**] (for the AIP on the Cisco 7500 series routers; for the ATM port adapter on the Cisco 7200 series)

show lane server [**interface atm** *slot/port-adapter/port*[*.subinterface-number*] | **name** *elan-name*] [**brief**] (for the ATM port adapter on the Cisco 7500 series routers)

show lane server [**interface atm** *number*[*.subinterface-number*] | **name** *elan-name*] [**brief**] (for the Cisco 4500 and 4700 routers)

Syntax	*Description*
interface atm *slot/port*	(Optional) ATM interface slot and port for the following: • AIP on the Cisco 7500 series routers. • ATM port adapter on the Cisco 7200 series routers.
interface atm *slot/port-adapter/port*	(Optional) ATM interface slot, port adapter, and port number for the ATM port adapter on the Cisco 7500 series routers.

Syntax	*Description*
interface atm *number*	(Optional) ATM interface number for the NPM on the Cisco 4500 or 4700 routers.
.subinterface-number	(Optional) Subinterface number.
name *elan-name*	(Optional) Name of emulated LAN. The maximum length of the name is 32 characters.
brief	(Optional) Keyword used to display the brief subset of available information.

Command Mode

EXEC

Usage Guidelines

This command first appeared in Cisco IOS Release 11.0.

Sample Displays

The following is sample output of the **show lane server** command for an Ethernet-emulated LAN:

```
Router# show lane server

LE Server ATM2/0.2  ELAN name: elan2  Admin: up  State: operational
type: ethernet          Max Frame Size: 1516
ATM address: 39.020304050607080910111213.00000CA05B41.02
LECS used: 39.020304050607080910111213.00000CA05B43.00 connected, vcd 51
control distribute: vcd 57, 2 members, 2 packets

proxy/ (ST: Init, Conn, Waiting, Adding, Joined, Operational, Reject, Term)
lecid ST vcd    pkts Hardware Addr  ATM Address
   1  O   54       2 0000.0ca0.5b40 39.020304050607080910111213.00000CA05B40.02
   2  O   81       2 0060.2f55.7940 39.020304050607080910111213.00602F557940.02
```

The following is sample output of the **show lane server** command for a Token Ring-emulated LAN:

```
Router# show lane server

LE Server ATM3/0.1  ELAN name: anubis  Admin: up  State: operational
type: token ring          Max Frame Size: 4544      Segment ID: 2500
ATM address: 47.00918100000000000000000.00000CA01661.01
LECS used: 47.00918100000000000000000.00000CA01663.00 connected, vcd 6
control distribute: vcd 10, 2 members, 4 packets
proxy/ (ST: Init, Conn, Waiting, Adding, Joined, Operational, Reject, Term)
lecid ST vcd    pkts Hardware Addr  ATM Address
```

```
   1  0    7       3 400.1          47.00918100000000000000000.00000CA01660.01
                     0000.0ca0.1660 47.00918100000000000000000.00000CA01660.01
   2  0   16       3 300.1          47.00918100000000000000000.00000CA04960.01
                     0000.0ca0.4960 47.00918100000000000000000.00000CA04960.01
```

Table 9–8 describes significant fields shown in the displays.

Table 9–8 *Show LANE Server Field Descriptions*

Field	Description
LE Server ATM2/0.2	Interface and subinterface of this server.
ELAN name	Name of the emulated LAN.
Admin	Administrative state, either up or down.
State	Status of this LANE server. Possible states for a LANE server include down, waiting_ILMI, waiting_listen, up_not_ registered, operational, and terminating.
type	Type of emulated LAN.
Max Frame Size	Maximum frame size (in bytes) on this type of emulated LAN.
Segment ID	The emulated LAN's ring number. This field appears only for Token Ring LANE.
ATM address	ATM address of this LANE server.
LECS used	ATM address of the LANE configuration server being used. This line also shows the current state of the connection between the LANE server and the LANE configuration server and the virtual circuit descriptor of the circuit connecting them.
control distribute	Virtual circuit descriptor of the Control Distribute VCC.
proxy	Status of the LANE client at the other end of the Control Distribute VCC.
lecid	Identifier for the LANE client at the other end of the Control Distribute VCC.
ST	Status of the LANE client at the other end of the Control Distribute VCC. Possible states are Init, Conn, Waiting, Adding, Joined, Operational, Reject, and Term

Table 9-8 *Show LANE Server Field Descriptions, Continued*

Field	Description
vcd	Virtual channel descriptor used to reach the LANE client.
pkts	Number of packets sent by the LANE server on the Control Distribute VCC to the LANE client.
Hardware Addr	The top number in this column is the router-descriptor, while the second number is the MAC-layer address of the LANE client.
ATM Address	ATM address of the LANE client.

Index

A

D

G

H

I-K

L

M

N

O

P-Q

R

S

T

U-V

W-Z

CISCO CERTIFIED INTERNETWORK EXPERT

Cisco's CCIE certification programs set the professional benchmark for internetworking expertise. CCIEs are recognized throughout the internetworking industry as being the most highly qualified of technical professionals. And, because the CCIE programs certify individuals—not companies—employers are guaranteed any CCIE with whom they work has met the same stringent qualifications as every other CCIE in the industry.

To ensure network performance and reliability in today's dynamic information systems arena, companies need internetworking professionals who have knowledge of both established and newer technologies. Acknowledging this need for specific expertise, Cisco has introduced three CCIE certification programs:

WAN Switching

ISP/Dial

Routing & Switching

CCIE certification requires a solid background in internetworking. The first step in obtaining CCIE certification is to pass a two-hour Qualification exam administered by Sylvan-Prometric. The final step in CCIE certification is a two-day, hands-on lab exam that pits the candidate against difficult build, break, and restore scenarios.

Just as training and instructional programs exist to help individuals prepare for the written exam, Cisco is pleased to announce its first CCIE Preparation Lab. The CCIE Preparation Lab is located at Wichita State University in Wichita Kansas, and is available to help prepare you for the final step toward CCIE status.

Cisco designed the CCIE Preparation Lab to assist CCIE candidates with the lab portion of the actual CCIE lab exam. The preparation lab at WSU emulates the conditions under which CCIE candidates are tested for their two-day CCIE Lab Examination. As almost any CCIE will corroborate, the lab exam is the most difficult element to pass for CCIE certification.

Registering for the lab is easy. Simply complete and fax the form located on the reverse side of this letter to WSU. For more information, please visit the WSU Web page at `www.engr.twsu.edu/cisco/` or Cisco's Web page at `www.cisco.com`.

CISCO CCIE PREPARATION LAB

REGISTRATION FORM

Please attach a business card or print the following information:

Name/Title: ____

Company: ____

Company Address: ____

City/State/Zip: ____

Country Code (_____) Area Code (_____) Daytime Phone Number ____________________

Country Code (_____) Area Code (_____) Evening Phone Number ____________________

Country Code (_____) Area Code (_____) Fax Number ____________________

E-mail Address: ____

Circle the number of days you want to reserve lab: 1 2 3 4 5

Week and/or date(s) preferred (3 choices):

____

____

____

Have you taken and passed the written CCIE exam? Yes No

List any CISCO courses you have attended:

Registration fee: ______________ $500.00 per day × _____ day(s) = Total ____________________

Check Enclosed (Payable to WSU Conference Office)

Charge to: ____________________________ MasterCard or Visa exp. Date ________

CC# ____

Name on Card ____

Cardholder Signature ____

Refunds/Cancellations: The full registration fee will be refunded if your cancellation is received at least 15 days prior to the first scheduled lab day.

Wichita State University
University Conferences
1845 Fairmount
Wichita, KS 67260
Attn: Kimberly Moore
Tel: 800-550-1306
Fax: 316-686-6520